OXFORD MEDICAL PUBLICATIONS

Male Victims of Sexual Assault

Male Victims of Sexual Assault

Edited by

Gillian C. Mezey

*Consultant and Honorary Senior Lecturer
Department of Forensic Psychiatry
St George's Hospital
London, UK*

and

Michael B. King

*Senior Lecturer and Honorary Consultant
Academic Department of Psychiatry
Royal Free Hospital School of Medicine
London, UK*

HV
6569
C7
M35
1992

Oxford New York Tokyo
Oxford University Press

Oxford University Press, Walton Street, Oxford OX2 6DP
Oxford New York Toronto
Delhi Bombay Calcutta Madras Karachi
Kuala Lumpur Singapore Hong Kong Tokyo
Nairobi Dar es Salaam Cape Town
Melbourne Auckland Madrid
and associated companies in
Berlin Ibadan

Oxford is a trade mark of Oxford University Press

Published in the United States
by Oxford University Press Inc., New York

© The contributors listed on p. ix, 1992

First published 1992
First published in paperback 1993

All rights reserved. No part of this publication may be reproduced,
stored in a retrieval system, or transmitted, in any form or by any means,
without the prior permission of Oxford University Press.
Within the UK, exceptions are allowed in respect of any
fair dealing for the purpose of research or private study, or criticism or
review, as permitted under the Copyright, Designs and Patents Act, 1988, or
in the case of reprographic reproduction in accordance with the terms of
licences issued by the Copyright Licensing Agency. Enquiries concerning
reproduction outside those terms and in other countries should be sent to
the Rights Department, Oxford University Press, at the address above.

This book is sold subject to the condition that it shall not, by way
of trade or otherwise, be lent, re-sold, hired out, or otherwise circulated
without the publisher's prior consent in any form of binding or cover
other than that in which it is published and without a similar condition
including this condition being imposed on the subsequent purchaser

A catalogue record for this book is available from the British Library

Library of Congress Cataloging in Publication Data
Male victims of sexual assault / edited by Gillian C. Mezey and
Michael B. King.
Includes bibliographical references.
1. Male rape victims—Great Britain—Psychology. 2. Male rape—
—Great Britain—Psychological aspects. 3. Rape trauma syndrome.
I. Mezey, Gillian C. II. King, Michael B.
[DNLM: 1. Men—psychology. 2. Sex Offenses. 3. Sex Offenses—
—psychology. W 795 M245]
HV6569.G7M35 1992   362.88'3'081—dc20   92-8441
ISBN 0 19 261871 7 (hbk)
ISBN 0 19 262469 5 (pbk)

Printed and bound in Great Britain by
Biddles Ltd, Guildford and King's Lynn

Preface

In the debate on sexual assault, men are portrayed as the perpetrators. Rarely is it appreciated that they may themselves be the victims of sexual abuse. There are three principal reasons for our inability to recognize the vulnerability of men. First, the male sexual stereotype emphasizes their superior strength, physical size, and role as initiator of sexual activity. Second, there is a failure to appreciate the nature of sexual assault as primarily an aggressive act, rather than one which is motivated by sexual need. Third, rape is narrowly defined in British law as non-consensual vaginal penetration by a penis. Thus, the gender of both perpetrator and victim is predetermined. Other forms of violation, such as anal or oral penetration of either sex, are relegated to the offences of indecent assault and buggery, which carry less severe penalties.

This failure of recognition is reflected in the lack of information about male sexual assault, its prevalence, characteristics, and sequelae. Nevertheless, it is apparent that male victims experience the act as one of violence and humiliation, which threatens their personal and physical integrity. The similarities in the responses of male and female victims to sexual assault are greater than the differences. However, there are issues specific to each sex that reflect cultural and sexual stereotypes and influence the victim's recovery and readjustment. These are highlighted by the experiences of men sexually assaulted in the community, where the circumstances of the assault parallel those of rapes reported by women.

In Chapter 1, by King, the results of a survey of 22 men assaulted in the community illustrate the assault's profound impact on them and the difficulty they may experience in coming to terms with it. In Chapter 2, West challenges the prevailing stereotype of homosexuals as perpetrators of attacks on heterosexual men. Sexual attacks on men may express antipathy towards, or fear of, homosexuality as well as the perpetrator's own underlying sexual conflicts. With an increasing awareness of the nature and extent of child sexual abuse comes the recognition that boys and male adolescents are frequently victimized. In Chapter 3, Watkins and Bentovim give a comprehensive overview of this area, including an examination of the links between early victimization and later sexual offending. Homosexual activity in institutions has long been considered as an unsatisfactory substitute for heterosexual relations, necessitated by an all-male environment. The coercive and violent nature of many sexual encounters in institutions, such as prisons, is discussed in Chapter 4 by King. The response of men to sexual assault is very similar to post-traumatic stress disorder that follows other life-threatening events. In Chapter 5, Turner describes reactions to torture to illustrate the mechanisms by which overwhelming experiences are processed and resolved. Silverman reminds us in Chapter 6 that friends and family of the victim may experience distress and, by their reactions, influence the victim's recovery. The historical and anthropological background to male sexual assault is discussed by Jones in Chapter 7. The damage to the victim is determined as much by the cultural context in which the assault occurs as by the traumagenic nature of the act itself. Adler, in Chapter 8, emphasizes the legal difference for men and women. For women, the law stands to protect the innocent man from false accusation, whereas much law relating to male sexuality stands to protect society from deviation from the traditional male heterosexual stereotype. Finally, men and women have an equal right to care and concern after an assault. Treatment issues, considered in the final chapter, by Mezey, include what type of help is required, who should provide that help, and how effective are current treatments in assisting victims to make a full recovery.

In this book we attempt to open the debate on male sexual assault, bringing together existing data, theoretical perspectives, and implications for future practice and policy.

London G.C.M.
January 1992 M.B.K.

Contents

List of contributors ix

1. Male sexual assault in the community
   Michael B. King 1

2. Homophobia: covert and overt
   D. J. West 13

3. Male children and adolescents as victims:
   a review of current knowledge
   Bill Watkins and Arnon Bentovim 27

4. Male rape in institutional settings
   Michael B. King 67

5. Surviving sexual assault and sexual torture
   Stuart Turner 75

6. Male co-survivors: the shared trauma of rape
   Daniel C. Silverman 87

7. Cultural and historical aspects of male
   sexual assault
   Ivor H. Jones 104

8. Male victims of sexual assault – legal issues
   Zsuzsanna Adler 116

9. Treatment for male victims of rape
   Gillian C. Mezey 131

Index 145

Contributors

Zsuzsanna Adler Police Staff College, Bramshill House, near Basingstoke, Hampshire RG27 0JW, United Kingdom.

Arnon Bentovim Hospitals for Sick Children, Great Ormond Street, and the Tavistock Clinic, London, United Kingdom.

Ivor H. Jones Department of Psychiatry, Clinical School, Royal Hobart Hospital, 43 Collins Street, Hobart, Tasmania 7000, Australia.

Michael B. King Academic Department of Psychiatry, Royal Free Hospital School of Medicine, Pond Street, London NW3 2QG, United Kingdom.

Gillian C. Mezey Department of Forensic Psychiatry, St George's Hospital, Blackshaw Road, London, SW17 0QT, United Kingdom.

Daniel C. Silverman 330 Brookline Ave, Boston, Massachussetts 02215, United States.

Stuart Turner Department of Psychiatry, Wolfson Building, Middlesex Hospital, Mortimer Street, London W1N 8AA, United Kingdom.

Bill Watkins Department of Psychological Medicine, Christchurch School of Medicine, Christchurch Hospital, Christchurch, New Zealand.

D. J. West Institute of Criminology, 7 West Road, Cambridge, United Kingdom.

1

Male sexual assault in the community

Michael B. King

Sexual assault of adult men is recognized as a problem in prisons Sagarin 1976; Anderson 1983) and other all male institutions (Goyer and Eddleman 1984). Until recently, however, there has been little consideration of sexual assault of men in the community. Statistics on male sexual assault are rare, encouraging disbelief in the phenomenon. This is due, in part, to the narrow legal definition of rape in the United Kingdom and in some states of the United States. In English law the term 'rape' is restricted to forced penile penetration of the vagina and thus cannot apply to sexual assaults against men. Forced anal penetration of a man is considered within the Sexual Offences (Amendment) Act of 1976 as non-consensual buggery and carries a lesser penalty.

At least 39 states in North America now have gender-neutral statutory rape laws (Sarrel and Masters 1982) and thus some estimate of the prevalence of male sexual assault is possible. Rape of men reported to law enforcement agencies contributes between 5–10 per cent to the total rapes reported (Forman 1982). In the United States there has also been a generalized finding of an increase in male victims among all rape cases, one study published in the late 1970s reporting an increase from 0 per cent to 10 per cent in three years (Kaufman *et al.* 1980). It must be emphasized that these figures concern incidents of rape *reported* to counselling or law enforcement agencies and, as is the case for women (Sutherland and Scherl 1970; Divasto *et al.* 1984; Howard League Working Party 1985), are probably a large underestimate of the true incidence.

Even when the definition of rape is widened to include any person without specification of gender, male rape is generally assumed to consist of assaults by homosexuals against heterosexual men and boys. This arises from the popular view of rape as a sexually motivated crime and the assumption that, if men are targets, their attackers must be of homosexual orientation (Groth and Burgess 1980). In recent articles concerning possible sexually transmitted diseases in rape victims, it has been argued that while a human immunodeficiency virus (HIV) test is particularly necessary for male victims (Osterholm et al. 1987, Hillman et al. 1990), it may not be necessary, in most cases, for women (Jenny et al. 1990). While it is certainly true that, in suffering more penetrative trauma, male rape victims are at greater potential risk for HIV transmission, a hidden assumption may arise in that, if men are raped at all, then their assailants must be homosexual and thus more likely to be HIV antibody-positive. Interestingly, evidence from the United States indicates that the reverse is true, with male rape more likely to be committed by heterosexual men against homosexual victims (Groth and Burgess 1980; Anderson 1982). Although so-called sexual harassment of men by women is claimed to occur more often than was previously assumed (Struckman-Johnson 1988), serious sexual assault of men by women is rare (Johnson and Shrier 1987) and occurs most often when the woman has a distinctive emotional or physical dominance over the man she attacks (Sarrel and Masters 1982; Masters 1986).

A survey

Although there has been increasing concern in North America about the numbers of male victims of sexual assault (Groth and Burgess 1980; Kaufman et al. 1980; Anderson 1982; Forman 1982; Goyer and Eddleman 1984; Kaszniak et al. 1988; Myers 1989), reports have been of small and atypical series of men often selected when presenting to emergency rooms or seeking psychiatric help. Prior to 1986 no work had been conducted in Britain into this problem.

The aim of the research to be described in this chapter was to study the context of male sexual assault in Britain by recording the experiences of victims and investigating the long-term psychological and social effects of such assaults. As the purpose was to investigate the nature and extent of sexual assault among men in the community, no attempt was made to ascertain subjects from medical or psychiatric clinics, or legal or social agencies. Populations ascertained in this manner may not be representative of all men who have been assaulted.

Bypassing such agencies made it necessary to advertise the study to encourage men who had been assaulted to come forward. Even by this method it remains possible that more of those having suffered serious assaults might perceive the research as salient and respond to a call to take part. By seeking people through an article in which were outlined the sorts of reactions men might suffer after a sexual assault, there was also a risk of inadvertently influencing their later responses to questions. Although not ideal, this was the only venue through which sufficient numbers of subjects in the wider community could be reached.

Several national daily newspapers and gay periodicals in the United Kingdom were asked to publish reports about the proposed research, including in their article a call for men with a history of sexual assault to take part. Editorial reaction was initially one of surprise and, not uncommonly, one of distaste. Although reporters were prepared to write about sexual assault on men, editors more senior in the publishing hierarchy were less enthusiastic and often vetoed an article before it went to press. Their grounds for preventing publication were based partly in disbelief and partly in embarrassment at the nature of the topic. Despite this reluctance, one national newspaper finally produced a sensitive report, enabling the research to begin. Members of the gay Press were less apprehensive of the reaction of their readers.

Men with a history of assault when they were aged at least 16 years were asked to make contact in order to receive a questionnaire concerning this assault. Each subject gave written consent to take part and confidentiality was stressed throughout. Demographic details of the victim and, where known, the assailant were sought before more sensitive questioning about previous victimization, sexual orientation of victim and assailant(s), circumstances and nature of the assault, prior relationship between victim and assailant, reactions in the short and long term, help seeking, use of alcohol and drugs, and, lastly psychiatric treatment, both before and after the attack. The questionnaire was semi-structured but with enough space for additional comment.

Each subject was also asked if he would be willing to take part in a semi-structured interview designed to collect a full personal history as well as more detailed information about the assault and its effects.

Results

Between 30 and 40 contacts were received from throughout the United Kingdom (Mezey and King 1989). Several were from people who were aware of friends who had been sexual assaulted or from interested organizations. Of 29 men who claimed to have been sexually assaulted, one wrote anonymously and three were under the age of 16 at the time of the assault. Questionnaires were sent to the remainder, of whom 22 eventually returned them. Four men did not acknowledge the request for interview, one (an elderly man) died before interview was possible, six refused and 11 agreed. Of the 11 respondents willing to take part in the interview, only eight eventually attended.

The subjects

All were caucasian and 15 were aged between 18 and 36 years. Mean age at the time of the attack was 26.3 years (range 16–82 years). All respondents lived in cities or towns. Three were students and 15 of the remaining 19 were in social classes 2 and 3 (Goldthorpe and Hope 1974).

Ten men were self-identified as homosexual, four as bisexual, and eight as heterosexual at the time of the assault. Four subjects, all of whom were heterosexual, reported no sexual experience prior to the assault.

Four homosexual men reported that they had suffered previous sexual victimization, two heterosexual men that they had been mugged, and two heterosexual men that they had experienced both sexual and personal assaults in the past.

The assailants

As this information was given by the victims it was often incomplete. Eleven men believed their attackers were homosexual, three that they were heterosexual, and three bisexual. In the remaining five cases, including one respondent who had been assaulted by several men, the sexuality of the assailants was not established.

Four men (two homosexual and two heterosexual) were attacked by complete strangers. Of the remainder, six were assaulted by someone well known to them, five by brief (a few hours) acquaintances, three by someone met for casual sex, three by homosexual lovers or ex-lovers, and one by a member of the family.

In seven cases the assailant held a degree of emotional or more formalized authority over the victim. For example, one heterosexual man was assaulted by a priest whom he viewed as a trusted confidant, another bisexual married man was attacked by a man who advertised himself as a counsellor for married gay men in a well-known national periodical, and a third man was assaulted by an Army officer of higher rank.

Thirteen men reported that their attackers had been drinking alcohol around the time of the assault, some of whom were heavily intoxicated. Alcohol use was suspected in a further five cases.

The assault

There was no particular pattern to the timing of the assaults over 24 hours or throughout the week. In nine instances the assault took place in the assailant's home, while in five it occurred in that of the victim. Although only six assaults happened outdoors, in all of these the assailants were completely unknown to the victims or were very new acquaintances (1–2 hours) and in every case the victims were injured, at least to some extent.

Forced anal intercourse was the principal assault on 17 men; in a further three, anal intercourse was attempted. Of the remaining two men, one was forced to perform fellatio and the second was indecently assaulted as part of a fierce physical attack. Eleven men were subjected to other forms of assault in addition to the buggery or attempted buggery, which usually included forced fellatio and (in two cases) being urinated upon. Five assailants attempted to masturbate their victims, three of whom succeeded in making them ejaculate. The men who were stimulated in this way were disgusted and confused by their physiological response:

I was angry and embarrassed, but frightened because the whole episode was like fantasy and reality getting mixed up. The fear was to do with my sexual response to pain.

Assailants were described by their victims as angry, scornful or sadistic in 14 assaults, and several men reported being verbally and physically humiliated. Most subjects experienced intense emotions of fear, unreality, anger, or revulsion during the attack, and 12 men believed they were about to be killed by their attacker. These feelings of fear and unreality overwhelmed them to such an extent that many were unable to mount any effective resistance:

He was much bigger. It was pure fright. I just wanted to protect myself. There was no way I could run out of the room. He had locked the door. He was too big – 6ft 1in or 6ft 2in. I thought if I gave in, it would be over quicker.

One claimed he was too drunk to resist, and another was assaulted so swiftly and brutally that no resistance was possible. Nevertheless, the remainder used at least verbal persuasion to escape the situation and nine men fought back physically, three of whom managed to escape, preventing forced anal intercourse, which was attempted by the assailants.

Immediate reactions

Although all of the men claimed that the assault had a major effect on their lives, only nine reported it to any other person in the immediate aftermath, and six disclosed the assault for the first time by responding to this research. Most were embarrassed and humiliated and feared they would not be believed:

All men find rape difficult to believe or accept – if you let it happen you must be queer, if you're not queer it can't have happened.

All of the men had felt the stigma or frank disbelief that

accompanies such assaults. One man disclosed that even on the day of the interview, he wondered if he would be taken seriously. Many had difficulty believing what had happened to them, particularly why they had been so fearful and unable to escape.

Although five men sought medical attention for their injuries, failure to seek help did not always reflect the seriousness of the assault. One man who was frightened by loss of blood clots from the rectum after the attack, nevertheless did not seek medical help out of embarrassment and a fear he might have to explain his injuries.

Few men considered reporting to the police was a serious option. Most homosexual victims were wary of the police, believing that they would be perceived as 'asking for it', and heterosexuals feared the humiliation and suspicion that they must be gay. A heterosexual man said: 'I had heard the police were very unsympathetic to female victims and couldn't imagine their reaction to a male rape.' A homosexual man said: 'I didn't think they [police] would take me seriously as I am homosexual. Police are basically anti-gay, let alone gay rape victims!'

None the less, the two men who did report to the police were managed appropriately and in both cases the assailant was apprehended and a court case resulted. One was homosexual and managed to hide this fact despite being challenged both by the police and by the courts. Both victims, however, were angry and disappointed that prison sentences imposed on their assailants were suspended.

Longer-term reactions

Only two men claimed to have suffered no psychological after-effects of the assault and appeared to have coped without major overt distress. Among the remainder there was a wide range of long-term sequelae in the form of a greater sense of vulnerability, increased anger and irritability, loss of self-respect, and rape-related phobias. Several victims confined themselves to their homes for some time afterwards and withdrew from friends and family. A heterosexual victim said:

My social life ended... my friends accepted this counterclaim that I had attempted to seduce him and made false accusations when rebuffed. They treated me as if I had contracted leprosy.

Several felt dirty and damaged and wondered if this was obvious to others around them:

I became extremely wary of everybody for a while and got the impression that everybody knew what had happened to me and was staring at me with a sense of disgust and contempt.

These reactions were sometimes experienced for years after the event:

I still have memories, usually triggered by the word rape or [the name of the suburb where it happened]. I think about it at least once a day, still.

Five of the eight heterosexual victims sought psychological treatment at some time after the assault, none of whom had had psychiatric help before this. Six homosexual men also sought psychotherapeutic help at some stage after the assault, although three had also had treatment beforehand. Not all of those seeking help after the assault were certain that there was a direct connection. Only two subjects revealed the attack to their psychiatrist and neither received a sympathetic response. One heterosexual man was: '... politely disbelieved and urged to come to terms with the homosexual side of myself'.

Although only one of the eight men who were interviewed had significant psychiatric symptoms at that time, two had attempted suicide after the assault. A further man committed suicide between completing the questionnaire and presenting for an interview. He was receiving counselling from a local victim-support scheme and reportedly left a note stating that the assault had played a part in the decision to end his life.

Use of alcohol or other drugs also changed as a consequence of the attack. Eight victims began to drink much more heavily, three of whom also increased their use of prescribed psychotropic drugs. Although this drug taking was usually temporary, one man later needed hospital treatment to withdraw from illicit drugs.

Sexual orientation

Although seven of the nine heterosexual men considered that the attack had had no long-term effect on their sexual orientation, several had wondered if there was a homosexual component to their personality that could have 'attracted' the assailant. Two heterosexuals later became ambivalent about their sexual identity but only one considered this a consequence of the assault.

Those four men (all were heterosexual) for whom the attack had been their first sexual experience had the greatest difficulty adjusting to later sexual relationships. One experienced problems because, in his words:'One fear was ... that I might make someone do something against their will, that is become an unintentional rapist.' Two others appeared not to have had any sexual contact since the attack and the fourth was confused as to his sexual orientation. Not uncommonly, the homosexual men experienced a resurgence of ambivalent feelings about their sexual orientation:

I had always felt that I had come to terms with my homosexuality but it caused serious problems for me and I felt that I had in some way lost my sexual identity. Sex became dirty.

Two men who were bisexual or homosexual at the time of the attack later went thought a period of promiscuity. The bisexual man said:

The attack made me feel really base – the lowest form of human life – it was the catalyst to my marriage break-up as heterosexual relations ceased and I became homosexually promiscuous.

Discussion

There are five limitations to this work. First, as discussed earlier, seeking responders through the media, although not ideal, was the only method possible for a preliminary study. Second, although strict privacy of all information was emphasized, there is little doubt that many men would fear that their accounts might become identifiable in some way and decline to come forward. Third, the fact that the work was conducted from a psychiatric research institute may have been a inhibiting factor for homosexual men for whom psychiatry does not have a particularly good record, and for heterosexuals the prospect of psychiatric contact, even for the purposes of research, may have made them feel more stigmatized. Fourth, many victims might have been reluctant to relive painful memories solely for purposes of research where there was no overt offer of support or therapy. Finally, although the questionnaire was carefully constructed, it is inevitable that much of the detail of the experiences and feelings of subjects will have been lost. Although interviews were planned to counteract this, as predicted, only a proportion were prepared to take part.

Despite these limitations, the results of this study demonstrate that male sexual assault is a frightening, dehumanizing event, leaving men who have been assaulted feeling debased and contaminated, their sense of autonomy and personal invulnerability shattered. These effects were most devastating when the men were sexually inexperienced before the assault.

There was no evidence that these men had a higher likelihood than women who are attacked of being beaten or sexually assaulted by multiple assailants, as has been described in the United States (Groth and Burgess 1980). Although it is assumed that adult men are able to defend themselves, these findings correlate with others (Anderson 1982) in indicating that, like women, men often react to extreme personal threat with frozen helplessness. Assailants sometimes exerted a psychological dominance in order to overpower their victim, a phenomenon also reported in assaults taking place in institutions such as the armed forces (Goyer and Eddleman 1984) The figures reported here do not help to establish how often men successfully repel an assailant. Men who had experienced an attempted assault, but had successfully escaped, might have been less likely to come forward in response to a request for *victims*. Nevertheless, the most common response reported by these subjects was one of helplessness and passive submission, engendered by an overwhelming sense of fear and disbelief. Submission in the face of a major threat is a primitive response for self-preservation (Storr 1968; Symonds 1975). It creates enormous problems, however, for victims' later resolution of the attack and their role in it. There is good evidence that juries are more likely to believe a woman's account of rape if there is testimony of physical injuries as evidence of an effective struggle (Adler 1987), and it is easy to see how much more this would apply to men.

Again, in contrast to American reports (Groth and Burgess 1984), more of our assailants were homosexual than heterosexual, although this finding was dependent on the accounts of the men assaulted. Homosexual or bisexual men were also more common among the victims. Possibly this arose as a selection bias, as we had advertised in the gay press. Some of the homosexual men may also have put themselves at risk when seeking out sexual partners. Homosexual meeting-places are often used by those intent on assaulting homosexuals and it was clear in some instances that 'queer bashing' of this type had occurred (See Chapter 2). Men who have been attacked in these settings often feel guilty and responsible for the assault taking place, in a similar manner to women whose assaults arise from meeting men in social situations. It is already well known that a woman's evidence is quickly discounted if any question mark can be raised about her 'sexual availability' (Adler 1987).

These attacks had a considerable impact on victims' sexual activity and were often associated with a need for considerable adjustment to later sexual activity, although there was little long-term effect on sexual orientation. Only one heterosexual man believed that his later homosexual feelings were in some way a consequence of the assault. This runs counter to an American report on the effects of enforced homosexual experiences for heterosexual men in prisons (Sagarin 1976), in which it was claimed that those men who were coerced and subdued into homosexual relationships continued the pattern, pursuing it in their post-prison years. The author regarded this as evidence of the 'malleability' of human sexual orientation but his conclusions are open to question in that only nine subjects were involved in the study, and those who were said to have continued to have homosexual relations were ascertained from homosexual social circles. Thus, there remains little clear evidence of what becomes of prisoners who have been compelled to have sex with other men over extended periods of time in conditions where any hope of rescue has been denied. Although another American report also emphasized the conflicts over sexual orientation and consequent homophobia resulting from sexual attacks (Myers 1989), the men described were selected on the basis of having sought psychiatric help and many had been abused as boys, thus this finding may not be typical of all men who are assaulted.

There were striking similarities between the reactions of male victims and those reported for women who have been sexually assaulted (Katz and Mazur 1979; Sutherland and Scherl 1970; Burgess and Holmstrom 1974; Mezey and Taylor 1988). The behavioural, somatic, and psychological components of the rape trauma syndrome as it occurs in women who have been assaulted (Burgess and Holmstrom 1974) have been well described. Shock and disbelief occurring at the time of and soon after the attack are followed by humiliation, embarrassment, self-blame, behavioural changes, and rape-related phobias before final resolution. Many of these features occurred with men who were assaulted. The stigma for men may be even greater, however, in a society which expects its male members to be self-sufficient physically and psychologically.

As is the case for female victims (Katz and Mazur 1979), only very few of these assaults conformed to the stereotype of a sudden, unprovoked attack by a complete stranger in a public place. Most often the attacker was known to the man, at least at the level of a brief acquaintance. There was also a marked lack of awareness on the part of many of these male victims. Most women might have been wary of getting themselves into 'compromising' situations. However, for men, there is no cultural expectation that they might be sexually assaulted, thus warning bells do not ring. This may be changing in the homosexual community where, as outlined in Chapter 2, there is a rapidly increasing awareness of the possibility of physical, with or without sexual, assault (Thompson et al. 1985).

Psychiatrists need to be aware of sexual assault as a possible antecedent of psychiatric symptoms in both men and women. They also need to be sensitive to patients' reluctance to disclose the information. We know that only a minority of female victims need formal psychiatric help. Most respond to sensitive counselling. Men, however, because of the added stigma and suspicion surrounding their assault, may be more likely to suffer longer-term reactions. Although this study is not large or representative enough to indicate the natural history of mens' reactions to sexual assault, the high proportion who needed psychiatric help at some point after the attack is striking.

The future

More definitive epidemiological work is urgently required to establish the extent of sexual assault among men in the community. There is also a need for legal revision to allow recognition of the seriousness of penetrative sexual attacks on both men and women that fall outside the current, narrow definitions of rape. Only greater recognition and understanding of the problem will reduce the stigma. Finally, there needs to be greater provision of informed counselling for men who have suffered such assaults. Although there are now at least two voluntary organizations in Britain which offer counselling to men, rape crisis centres for women and victim support schemes that counsel victims of crime will need assistance to widen their remit and make greater provision for men who have been sexually assaulted.

References

Adler, Z. (1987). *Rape on trial*. Routledge & Kegan Paul, London.
Anderson, C. L. (1982). Males as sexual assault victims: multiple levels of trauma. In *Homosexuality and psychotherapy*, pp. 145–63. Haworth Press,
Burgess, A. W. and Holmstram, L. L. (1974). Rape trauma syndrome. – *American Journal of Psychiatry*, **131**, 981–6.
Divasto, P. U. et al. (1984). The prevalence of sexually stressful events among females in the general population. *Archives of Sexual Behaviour*, 13, 59–67.
Forman, B. D. (1982). Reported male rape. *Victimology*, 7:235–6.
Goldthorpe, J. H. and Hope, K. (1974). *The social grading of occupations. A new approach and scale*. Clarendon Press, Oxford.
Goyer, P. F. and Eddleman, H. C. (1984). Same-sex rape of nonincarcerated men. *American Journal of Psychiatry*, **141**, 576–9.
Groth, A. N. and Burgess, A. W. (1980). Male rape: offenders and victims. *American Journal of Psychiatry*, **137**, 806–10.
Hillman, R. J., Tomlinson, D., McMillan, A., French, P. D., and Harris, J. R. W. (1990). Sexual assault of men: a series. *Genitourinary Medicine*, **66**, 247–50.
Howard League Working Party Report (1985). *Unlawful sex. Offences, victims and offenders in the criminal justice system of England and Wales*. Waterlow, London.
Jenny, C. J. et al. (1990). Sexually transmitted diseases in victims of rape. *New England Journal of Medicine*, **322**, 713–16.
Johnson, R. L. and Shrier, D. (1987). Past sexual victimization by females of male patients in an adolescent medicine clinic population. *American Journal of Psychiatry*, **144**, 650–2.
Kaszniak, A. W., Nussbaum, P. D., Berren, M. R., and Santiago, J. (1988).

Amnesia as a consequence of male rape: a case report. *Journal of Abnormal Psychology*, **97**, 100–4.

Katz S. & Mazur, M. A. (1979). *Understanding the rape victim: a synthesis of research findings*. New York, John Wiley & Sons.

Kaufman, A., Divasto, P., Jackson, R., Voorhees, H., and Christy, J. (1980). Male rape victims: noninstitutionalised assault. *American Journal of Psychiatry*, **137**, 221–3.

Masters, W. H. (1986). Sexual dysfunction as an aftermath of sexual assault of men by women. *Journal of Sexual and Marital Therapy*, **12**, 35–45.

Mezey, G. and Taylor, P. J. (1988). Psychological reactions of women who have been raped. *British Journal of Psychiatry*, **152**, 330–9.

Mezey, G. and King, M. B. (1989). The effects of sexual assault on men: a survey of 22 victims. *Psychological Medicine*, **19**, 205–9.

Myers, M. F. (1989). Men sexually assaulted as adults and sexually abused as boys. *Archives of Sexual Behaviour*, **18**, 203–15.

Osterholm, M. T., MacDonald, K. L., Danila, R., and Henry, K. (1987). Sexually transmitted diseases in victims of sexual assault. *New England Journal of Medicine*, **316**, 1024.

Sagarin, E. (1976). Prison homosexuality and its effect on post-prison sexual behaviour. *Psychiatry*, **39**, 245–57.

Sarrel, P. M. and Masters, W. H. (1982). Sexual molestation of men by women. *Archives of Sexual Behaviour*, **11**, 117–31.

Starr, A. (1968). *Human Aggression*, Allan Lane, Penguin Press, England.

Struckman-Johnson, C. (1988). Forced sex on dates: it happens to men, too. *The Journal of Sex Research*, **24**, 234–41.

Sutherland, S. and Scherl, D. J. (1970). Patterns of response among victims of rape. *American Journal of Orthopsychiatry*, **40**, 503–11.

Symonds, M. (1975). Victims of violence: psychological effects and after effects *American Journal of Psychoanalysis*, **35**, 19–26.

Thompson, N. L., West, D. J., and Woodhouse, T. (1985). Social and legal problems of homosexuals in Britain. In *Sexual Victimisation*, (ed. D. J. West), pp. 93–159. Aldershot, Gower.

2
Homophobia: covert and overt
D. J. West

The concept of homophobia

Properly used, the word 'homophobia' refers to an intense, irrational fear of homosexuality, a pathological over-reaction presumably caused by intra-psychic conflict. Its clearest manifestation is the state of 'homosexual panic' that some men experience when another male suggests or attempts sexual contact. They may react with frenzied violence utterly disproportionate to the tentative or innocuous gesture that offended them. The phenomenon is thought to be due to unacceptable homosexual impulses inadequately repressed so that homosexual situations become peculiarly threatening to a self-image of pure heterosexuality. To relieve discomfort, the idea of any personal involvement is vigorously rejected and all responsibility is projected on to offending 'queers'.

Forensic psychiatrists are all too familiar with mitigating pleas put forward by men who, having murdered and robbed a homosexual, plead that they were driven to desperation by the victim's disgusting behaviour. Most masquerade as male prostitutes and accompany their victim home with the intention to extort, by threat or force, much more than the proposed price of their sexual favours. Resistance by the victim or fear of detection are the likely reasons for murder under these circumstances.

On rare occasions, attacks are precipitated by homosexual panic, as illustrated by the following case. The killer had battered to death a former employer whose homosexual interests were well known. He had helped the victim, who was drunk, to get home, but on arrival the drunken man made a sexual approach, which put him into a blind rage. There was no theft and an absence of any reasonable motive otherwise. There was, however, a history of several previous violent reactions to similar provocation. Once he had had to be restrained by bystanders as he was attacking a man said to have touched his thigh in a cinema. Although a husband, father, and heterosexual adulterer, he believed, for no adequate reason that could be seen, that his appearance caused people to think he was gay. In prison he was more preoccupied with worries about what others were thinking about his sexuality, in view of the nature of his crime, than he was about his forthcoming trial.

The word homophobia has passed into common usage and is applied more or less indiscriminately to any degree of dislike of homosexuality or homosexuals. The theory that such dislike is a projection of suppressed homosexual inclinations may have discouraged some people from showing their feelings too openly. As tolerance towards different sexual proclivities has increased among the liberal-minded this, too, may have brought about some muting of expressions of instinctive dislike. Even so, disapproval, both overt and covert, remains pervasive and reveals itself in many forms, not least in the way laws and social standards are applied differently to homosexual and heterosexual behaviour.

Origins of the taboo

Hostility towards groups or minorities whose habits and ideas differ from one's own is an almost universal human characteristic. Many words have been coined to describe varieties of the phenomenon – xenophobia, racism, sexism, snobbery, anti-Semitism, and so forth. Prejudice against gays is but one example. Love and attraction is such an important component of most people's lives that a minority whose feelings or behaviour in this regard differ sharply from that of the majority cannot but cause unease. Their very existence challenges fundamental assumptions about life's purpose and the natural order of things. There is really no need to invoke Freudian psychopathology to account for homophobia.

Condemnation of homosexual behaviour is deeply rooted in history and tradition. Orthodox Judeo-Christian teachings are unequivocally antagonistic. Like other ancient rules, such as those forbidding the eating of pork, the strictures against homosexuals in the book of Leviticus and elsewhere in the Bible may once have had some sound practical basis. When the path from conception to maturity was so uncertain, procreation was a duty essential to the preservation of the tribe. Behaviour that diverted sexual energy elsewhere – such as Onan's spilling of his seed – was antisocial. When women without husbands to support them had no means of livelihood, men disinclined to take a wife would create problems. Even today, the deliberately childless may find themselves regarded as failing to take their fair share of social responsibility.

Popular hostility towards homosexuals, especially those who proclaim their way of life to be a basic human right, continues to receive strong backing from religious leaders. Protestant and Muslim fundamentalists, as well as Roman Catholics, regardless of any practical justification, continue to pronounce on the wickedness of homosexual behaviour. In October 1985, Pope John Paul II reaffirmed Catholicism's absolute condemnation of homosexual acts and of the promotion of gay rights.

Pronouncements by medical authorities in the past have been influential in promoting homophobia by equating homosexuality with 'psychopathy', 'degeneracy', and more turpitude. For many years it was the received wisdom in medical circles that severe personality problems invariably accompanied a homosexual orientation. Although Freud (1951) is credited with having challenged this assumption in his famous 'Letter to an American mother', his followers were still insisting upon it many years later (Socarides 1968). It was only after much protest and lobbying on behalf of the gay liberation movement that the American Psychiatric Association (1980) voted to delete homosexuality *per se* from its diagnostic manual of psychiatric disorders.

Prior to the introduction of partial decriminalization by the England and Wales Sexual Offences Act 1967, homosexual behaviour between males was, under all circumstances, a serious crime, and it remained so in Scotland for some years after that and in Northern Ireland for even longer. Until the agitation for law reform began in England in the 1950s, homosexuality rarely featured in the Press, save for accounts of trials and convictions which were less than explicit about the nature of the sexual misconduct, and was not a topic of everyday conversation. In the absence of objective information, mythical ideas about the subject could flourish unchecked.

Myths and half-truths

Unpopular groups attract derogatory images. Jews have Shylock, the poor have 'coals in the bath', blacks have muggers, and Asians have child labour. Male homosexuals are regarded as child molesters, traitors, transvestites, and effeminates; lesbians as man-haters and chauvinist females. Doubtless some are, but that is not to say they are in any way typical or representative.

The majority of male homosexuals are not effeminate, even though it is true that boys identified as 'sissy' in their early years are more likely than others to become homosexual adults (Green 1987). However, with the growing fashion in gay circles of moustache-sporting 'clones', to say nothing of the leather and chains 'uniform', the original image is becoming blurred. Various degrees of gender dysphoria, up to and including transsexuals who believe themselves trapped in a body of the wrong sex, may occur among persons of homosexual orientation. In reality, however, the incidence is far less than might be expected from the popular stereotype of limp-wristed, lisping buffoons portrayed by comedians. Nevertheless, false images and half-truths are perpetuated and slang words with pejorative overtones, such as 'queer', 'bent', 'poofter', or 'dyke', creep into the language. Although analogous expressions used publicly about racial minorities would be illegal, homosexuals have no such protection.

An accusation more often levelled against male homosexuals than against lesbians is that they are dangerous to the young. The complaint takes two forms: child molestation and the corruption of youth. In this context, corruption presumably means introduction to the practice of homosexuality. Young men are considered vulnerable to proselytism, material inducements, or seductive approaches from established homosexuals. This was one of the arguments that resulted in the exclusion of members of the armed forces and the merchant navy, and all men under 21, from the decriminalizing provisions of the 1967 Act. More recently, similar concerns have emerged in England and Wales in the controversial Section 28 of the Local Government Act 1988. This prohibits local authorities from 'promoting homosexuality by teaching or by publishing material'. More specifically it bans 'the teaching in any maintained school of the acceptability of homosexuality as a pretended family relationship'. This curiously worded provision was put forward after an outcry in the media about a children's book that described, and by implication approved, the situation of a young girl being brought up by two gay men.

The existence of a supportive subculture offends the homophobic and may well encourage some people, who might otherwise remain hidden, to give less furtive expression to homosexual feelings. The claims of modern gays, that homosexual and heterosexual love is equally fulfilling and equally deserving of respect, are a far cry from 'the love that dare not speak its name'. Whether all this has much influence on the determination of sexual orientation is doubtful. Survey results suggest that the hard core of exclusive homosexuals is of similar size in societies with radically differing attitudes towards them (Whitam 1983). Men who prostitute themselves to other men, but who do not class themselves as homosexuals, seem able to pursue their trade indefinitely without losing their basic heterosexual drive. The personal histories of so-called primary homosexuals (those who have never been heterosexually arousable) suggest that their homosexual urges develop very early in life, before social pressures or seduction experiences come into operation.

The recent explosion of concern in the media and among social workers about child sexual abuse, both inside and outside the home, and the realization that boys as well as girls are often the target of indecent behaviour by adults (usually males), increases the already strong suspicion of a causal link between male homosexuality and child molestation. The truth is that most men who have an erotic fixation on children and most of those who associate with children, because they cannot obtain or cope with an adult partner, are primarily interested in girls, although some will try to make contact with either boys or girls. There is no evidence that homosexuals are more likely than heterosexuals to be attracted to prepubertal children, or of any association between homosexuality and paedophilia. Surveys using the penile plethysmograph to assess sexual arousal have found that most men show some slight arousal to pictures of nude children, but if anything heterosexuals more so than homosexuals (Freund 1981). Clinical experience suggests that older men who 'regress' from adult relationships to paedophile offending are very rarely homosexual (Groth and Birnbaum 1978; Newton 1978). Double standards with regard to the age of sexual partners in heterosexual and homosexual relationships contribute to the myth about homosexual predilection for children. In a heterosexual context, the popularity of youthful pin-ups and teenage prostitutes bears witness to the attractiveness of adolescents for many normal men. When a similar age group is involved in male homosexual relationships it appears somehow more reprehensible as well as being technically criminal.

The myth of the homosexual predator against youth is not without some roots in reality. Homosexual paedophiles most certainly exist and can be both obsessively persistent in their offending behaviour and able to argue in favour of their right to 'Greek love' so long as the boy is willing and benefits from the relationship (Brongersma 1986). In the gay subculture based on bars, clubs, and discos, frequented by men on the lookout for casual sex, a high valuation is placed on the bloom of youth. Contacts between more mature adults and teenagers, sometimes facilitated by the transfer of money or other material benefits, is a noticeable feature of this scene.

Suggestions that homosexuals are unreliable or unsuitable for positions of responsibility, especially for sensitive posts in government, is another myth. Homosexuals are, of course, present in every walk of life and work as effectively as any other person. Nevertheless, substantial sections of the public do not trust them. A national survey conducted in the United Kingdom (Airey 1984) revealed that about half the British population considered it unacceptable for homosexuals to be teachers or to hold responsible public positions, and over 90 per cent agreed that male homosexual couples should not be allowed to adopt children under the same conditions as other couples.

Homosexuals working abroad arouse suspicion because of their supposed vulnerability to entrapment in compromising circumstances, after which they can be blackmailed into spying for a foreign power. This may indeed have happened on occasions (Vassall 1975), but similar pressures can be brought to bear upon adulterous heterosexuals. Paranoid attitudes towards homosexuals working in government offices reached a peak in the United States in the McCarthy era, when many unfortunate people lost their jobs.

Of all the stigmatizing labels applied to homosexuals the worst is that of being carriers of AIDS. In Central African countries, where the epidemic has spread further than it has in most other places, males and females are affected equally and there is no particular association with homosexuality. The American male homosexual community was unfortunate in being the first in the Western world to encounter the virus and to be blamed for its spread. Infection is communicated through the exchange of semen or blood containing the virus, which occurs through sexual intercourse, vaginal or anal. Habits of sexual promiscuity, particularly in the specialized steam-baths and dark backrooms of gay bars and cinemas which once featured prominently in the American gay lifestyle, promoted rapid spread of the so-called gay plague.

Male homosexuality had long been known to be associated with a relatively high incidence of sexually transmitted diseases such as syphilis and hepatitis. The advent of AIDS has brought about a dramatic change in the behaviour of male homosexuals. American bath-houses and dark-rooms have been shut down, and the gay Press has exhorted its readership to reduce the number of casual sex partners and use condoms and safer methods of mutual stimulation. Because of the specialized readership it has been possible to be more explicit and more colloquial than in materials addressed to the wider heterosexual public, and this has probably increased its effectiveness. Because of the long incubation period of the disease, any change of habits takes time to produce an observable effect, but in San Francisco, where the problem of HIV infection was especially acute and health education sponsored by gay organizations has been going on for some years, there has been a very significant decrease in the prevalence of positive reactions to the HIV antibody test among the gay community (Winkelstein *et al.* 1988). There has also been a significant decrease in incidence of other forms of sexually transmitted disease. Heterosexuals have not changed their habits to such degree, possibly because they feel less at risk, and heterosexual victims are becoming more numerous. Many men and women have sexual contact with both sexes, and heterosexual intravenous drug-users often pick up the virus from shared needles and syringes containing residues of infected blood. HIV infection is clearly not just a gay problem.

Legal discrimination

Notwithstanding official decriminalization, English criminal law and practice retains a range of provisions that ensure stricter control and harsher treatment of transgressors where homosexual as opposed to heterosexual behaviour is in question. Far from standing for equal rights and non-discrimination, as some would wish, the law appears to favour homophobic attitudes.

For example, the offence of 'persistently importuning male persons for immoral purposes', commonly regarded as the male equivalent of female street soliciting for prostitution, attracts harsher penalties. Although female soliciting has been made a non-imprisonable offence, male importuning still carries a maximum penalty of two years' imprisonment. Women have to be given a police warning when first apprehended, not so men. Women commit an offence only if they are asking for money for sex, but men who are merely looking for a partner, with no intention of demanding payment, are still liable to conviction, the courts automatically assuming that all homosexual propositioning is covered by the phrase 'immoral purpose'. Female soliciting is a street offence, but men can be prosecuted for importuning in any situation, including the environs of gay bars. In London, men found 'loitering' are often dealt with on the catch-all charge of highway obstruction.

As the age of consent for male homosexual acts in Britain is 21, if either or both partners are under that age, even if both are adolescents, they commit a crime. Body contact or mutual masturbation under these circumstances counts as 'gross indecency' and attracts a maximum penalty of two years' imprisonment, or its equivalent in the case of juveniles. If consensual anal intercourse (buggery) occurs, the maximum penalty is raised to 10 years or to life imprisonment if a boy under 16 is involved. As consensual anal intercourse generally occurs in private between persons who have established a relationship, the effect of the law is to penalize such persons more heavily than the more predatory offenders who have numerous and more casual contacts with many different young people encountered in public places. Late adolescence is a time of life when the sexual appetite is at a peak but young male homosexuals who are sexually active are technically criminals.

If members of a heterosexual couple behave 'indecently' together in a place where they can be seen by and offend a member of the public, they commit a minor, common-law offence. A male pair doing the same commit a serious statutory crime, either of gross indecency or buggery, with the severe maximum penalties just described. Moreover, 'in public' has a wider connotation where homosexual behaviour is involved. If a third party is present in the same house when homosexual acts occur, this removes the decriminalizing protection of the 1967 Act. 'Threesomes' or 'orgies' are crimes if the men included have contact with each other.

The letter of the law is often less important than the manner of its application. In some areas, indecency in men's lavatories has been pursued with peculiar vigour by local police who watch through spyholes or conceal themselves in broom closets to catch the unwary. Many people believe the police would be more profitably occupied in other pursuits. Gays have also complained of provocatively dressed young policemen waiting around in likely situations, hoping to be propositioned so they can make an arrest for importuning. As this ploy amounts to near incitement, it no longer has official approval. Nevertheless *agents provocateurs* have a long history in British law for use in this context (Davenport-Hines 1990). Men who have reported to the police violent robberies or thefts by male guests they have invited home for the night sometimes complain of an unsympathetic reception and of greater interest being shown in the possibly illegal nature of their own sexual behaviour than in their victimization.

Where the shoe pinches

The niceties of the law or of religious dogma, even the nastier comments in the tabloid Press and the awareness of widespread public disapproval, do not necessarily impinge directly upon the private lives of many homosexuals. In a survey among men self-identified as gay, Thompson *et al.* (1985) found that many respondents appeared contented and problem-free. Their sample was predominantly well educated, middle class, and beyond the age when crises of sexual identity are likely to occur. All the same, there were more than enough examples of overt conflicts and unhappy confrontations to show that a homosexual orientation is still a source of difficulty in Britain. Among a subsample of 100 who were interviewed in detail, as many as 49 reported having been physically attacked at least once in connection with their homosexuality. Many of the attacks had taken place when the respondent was 'cruising' known picking-up places and attracted the attention of 'queer bashers'. Other attacks arose in casual sexual encounters with strangers whose attitudes or motives were misjudged, some being intent on robbery and others having ambivalent reactions following sexual contacts. A minority of attacks were by jealous lovers.

Although blackmail attempts were not uncommon, being reported by 15 of the 100 men interviewed, most were not regarded as very serious. Most were attempts by strangers exploiting compromising sexual situations to obtain money. Some were for reasons of jealousy rather than financial gain, and one was a man threatened by his own parents to make him stop seeing his lover.

Brushes with the police were also quite common. Of the total of 443 respondents, 10.4 per cent reported having been charged with some offence in connection with homosexuality, and a further quarter had been questioned by the police at some time but not charged. The majority of the charges were for gross indecency or importuning, usually in or around public lavatories or cruising areas. Some arrests were in connection with demonstrations and other political activist situations concerning the rights of homosexuals.

It might be said that homosexuals have only themselves to blame for frequenting public places, especially public lavatories, for sexual contacts. Psychologists have suggested that part of the attraction is the thrill of risks being taken. There are probably simpler explanations. Gay bars and other 'legitimate' meeting places are not available in small towns, under-age youths have no access to them, and men living with families or whose jobs would be at risk cannot afford to be seen in such places. Secretive encounters in public lavatories cost nothing and may be resorted to for lack of other outlets.

Violence, blackmail, and confrontation with the police are the more dramatic problems of a homosexual lifestyle, but they are not necessarily the most prevalent or the most worrying. Feelings of oppression, non-acceptance, being an outsider, or being subject to subtle discrimination were much the most commonly voiced complaints. Many felt that they had lost friends or that their career prospects had been affected for the worse by disclosure of their homosexuality, and in fact there was some discrepancy between the positions they had obtained and their educational attainments. Criticism from parents and siblings had been a frequent and often hurtful experience, although some had succeeded in becoming reconciled after the turmoil following first disclosure. Even in today's supposedly more tolerant and sophisticated society, some young people suffer acutely from parental rejection on account of their sexuality or from a perceived necessity constantly to conceal their feelings and disguise the nature of their relationships. This kind of stress, coupled with their own unhappy ambivalence about becoming 'one of them' and adopting a gay identity, is an important cause of youthful suicides. A survey conducted by the London Gay Teenage Group found that one in five young homosexuals, both male and female, had at one time or other attempted suicide (Trenchard and Warren 1984). A common story from young men who have taken to prostitution is that they did so in order to survive after running away from or being turned out of parental homes where their homosexuality was unacceptable.

Families who first discover a son's sexual orientation when he is dying of AIDS face an enormous emotional upheaval. A priest who was visiting a patient dying in hospital told me recently of what happened when the family found out that the young man had been cohabiting for some years with a lover a little older than himself. The patient's brother, a 'born again' Christian, called at the hospital, warned him he was going straight to hell, demanded that he tell the lover to cease bedside visiting, and told the lover to pack up all the man's belongings so that the family could arrange for them to be removed immediately from the flat they shared. At the same time he berated the priest for showing tolerance towards their wicked relationship.

The activities of fundamentalist religious groups can have a demoralizing effect on homosexuals trying to achieve self-respect. In the January 1989 issue of the evangelical paper *Challenge*, an English Baptist preacher, head of an organization called U-Turn Anglia, whose homosexual son committed suicide, set out his attitude on the sinfulness of homosexuality and the divine judgement of AIDS. His organization seeks to bring 'deliverance' from damnation to those willing to accept biblical revelations and renounce their gay lifestyle.

The disapproval of people who matter – parents, siblings, work colleagues, employers, or landlords – is a discouraging influence on the setting up of stable cohabitations, the absence of which contributes to loneliness, secretive promiscuity, or forced and often disastrous marriages. Those who do form stable unions are often cut off from their disapproving families. Lacking the protection of a legally recognized marriage contract, when one dies without having made an appropriate will, the partner of many years standing may find himself suddenly turned out of the home by strangers demanding their inheritance. People in exposed public positions, such as politicians, prominent entertainers, teachers, and priests, have good reason to try to lead a 'closeted' existence. In doing so, however, they may resort to paying for sexual release with prostitutes or others who are in a position to blackmail them. Over the years the media have exploited such scandals to the full, and individuals such as a one-time leader of the British Liberal Party, a private detective employed by the Queen, and several Members of Parliament have all been hounded out of office as a result.

The reality of discrimination is reflected in numerous reports in the gay Press of abusive incidents and in the existence of gay organizations such as the Gay Civil Liberties Trust, with the specific purpose of fighting back. The London Lesbian and Gay Switchboard runs a 24-hour help-line and maintains a list of recommended solicitors for gay people who have been arrested. One such organization, GALOP, The Gay London Policing Group, publishes an annual report detailing examples of attacks on gays and of what they consider insensitive, oppressive, or neglectful policing. Their 1986/7 report, for instance, mentions that one night in July 1986 a young man was stabbed to death only hours after a gay man, who had been chased and threatened by a gang of 'queer bashers', had made an emergency telephone call to the police and no action had been taken. A frequent complaint is of bullying tactics in the course of police inquiries when the persons questioned are suspected of homosexuality. The GALOP reports also mention and give credit for helpful and professionally proper police responses. GALOP's 1987/8 report takes up the issue of 'half guilty acquittals', where men accused of indecency are pressured into consenting to being 'bound over to keep the peace' in instances where the case against them is too weak to secure a conviction. Another example of an organization fighting back is the Stonewall Housing Association set up to help gays find accommodation, a vital service for the victims of eviction by disapproving landlords or parents.

Official attitudes

Now that it is well known that homosexuals comprise a significant minority of the population, and with the acknowledgement that minorities should have equal rights with other citizens, politicians have become more wary about making comments amounting to blanket condemnation. All the same, 'gay rights' is a generally unpopular and suspect issue. Outside particular pressure groups in Britain, such as the National Council for Civil Liberties, there is little sympathy for those wanting to promote legislation to protect homosexuals in the way that ethnic minorities are protected. Industrial tribunals have not been favourably disposed towards homosexuals trying to claim that their exclusion from particular posts was unreasonable (Crane 1982). In any event, employers wishing to be rid of a homosexual can usually find less controversial grounds for doing so.

Reform of the criminal law in Britain to abolish discrimination on such matters as the age of consent, although recommended by, for example, the Howard League (1985), does not have the backing of the government-appointed Criminal Law Revision Committee (1984) and is not an issue commanding widespread public support. Local governing authorities who tried to publicize non-discriminatory policies attracted harsh criticism and were labelled the 'looney Left' by members of central government and by the media. Their activities in support of homosexuals have now been obstructed by Section 28 (see p. 16), which prohibits all 'promotion' of homosexuality by local government authorities.

In the United States, where a written constitution guarantees citizens' rights, attempts have been made to induce the Supreme Court to review and declare unconstitutional the laws which, in half of the States, still define all homosexual behaviour as a crime. A test case from Georgia, arising out of a prosecution for consensual acts in private, led to a decision that there is no fundamental right to engage in sodomy and that the rights of privacy do not extend to homosexual behaviour (*Bowers* v. *Hardwick*, 106, Supreme Ct. 2841 – 1986). Likewise, attempts at local level to introduce a ban on discrimination on grounds of sexual orientation have led to bitter contests and been fairly ineffectual (Meeker *et al.* 1985).

Compared with continental Europe, the United Kingdom has lagged behind in matters of homosexual law reform. It was only as a result of a finding of the European Court in 1981 that the 1967 decriminalizing Act was extended to Northern Ireland. The successful appeal in question (*Dudgeon* v. *UK* 1981) was against a breach of Article 8 of the European Convention on Human Rights, which guarantees respect for private life. The applicant had been interrogated by the police about homosexual friendships after they had discovered incriminating evidence in letters found in his home during a search for prohibited drugs.

In the more recent judgement (October 1988) in the case of David Norris, an Irish senator, it was held that the maintenance in Eire of the law penalizing homosexual acts in private constituted a continuing interference with the applicant's private life, contrary to Article 8. The Court declared that 'Such justifications as there are for retaining the law in force unamended are outweighed by the detrimental effects which the very existence of the legislative provision in question can have on the life of a person of homosexual orientation . . .'. Eire is exceptional among European countries in having resisted the legalization of homosexual relationships.

On the wider issue of non-discrimination, the United Kingdom still trails most of Europe. In 1981, Norway was the first in Europe to introduce a law forbidding anyone to threaten publicly, or to hold in contempt a person or group, or to refuse the use of facilities, on the grounds of homosexual orientation or way of life. On 13 March 1984, the European Parliament passed a resolution urging member states to apply the same age of consent for heterosexual and homosexual acts and called upon the European Commission to produce proposals to prevent discrimination against homosexuals in access to employment. These developments have been ignored in the United Kingdom and given virtually no publicity in the media. The former Prime Minister Margaret Thatcher's well-known adherence to 'Victorian values', her condemnation of 'permissiveness', and her reluctance to be dictated to by the rest of Europe gave no encouragement to would-be reformers.

Conclusion

In view of the state of the law regarding homosexual behaviour in many parts of the world and the existence of the death penalty in some countries, anxiety and perhaps even a certain paranoia is understandable on the part of the homosexual minority. The persecution of homosexuals alongside Jews in Hitler's concentration camps is within the memory of many still living.

Although the status of homosexuals in British society remains somewhat ambiguous, this owes more to the adverse views of some vociferous sections of the public than to any legal provision. Persecution of homosexuals has no official backing in the United Kingdom. Nevertheless, we have still a long way to go in establishing a fully tolerant and humane society that will afford persons of minority sexual orientations the same respect that heterosexuals enjoy.

References

Airey, C. (1984). Social and moral values. In *British social attitudes. The 1984 report*, (ed. R. Jowell and C. Airey). Social and Community Planning Research/Gower, Aldershot.

American Psychiatric Association (1980). *Diagnostic and statistical manual of mental disorders* (3rd ed). American Psychiatric Association, Washington DC.

Brongersma, E. (1986). *Loving boys*, Vol. 1. Global Academic, Elmhurst, NY.

Crane, P. (1982). *Gays and the law*. Pluto Press, London.

Criminal Law Revision Committee (1984). *Fifteenth Report: Sexual offences*. HMSO, London.

Davenport-Hines, R. T (1990). *Sex, death and punishment. Attitudes to sex and sexuality in Britain since the Renaissance*. Collins, London.

Freud, S. (1951). Letter to an American Mother. *American Journal of Psychiatry*, **107**, 252.

Freund, K. (1981). Assessment of paedophilia. In *Adult sexual interest in children*, (ed. M. Cook and K. Howells). Academic Press, London.

Green, R. (1987). *The sissy boy syndrome and the development of homosexuality.* Yale University Press, Newhaven, CT.

Groth, A. N. and Birnbaum, H. J. (1978). Adult sexual orientation and attraction to underage persons. *Archives of Sexual Behaviour,* **7**, 175–81.

Howard League (1985). Offences, victims and offenders in the criminal justice system of England and Wales. *The report of the Howard league working party.* Waterlow Publishers Ltd., London.

Meeker, J. W., Dombrink, J., and Geis, G. (1985). State law and local ordinances in California barring discrimination on the basis of sexual orientation. *Dayton Law review,* **10**, 745–65.

Newton, D. E. (1978). Homosexual behaviour and child molestation: A review of the evidence. *Adolescence,* **13**, 29–54.

Socarides, C. W. (1968). *The overt homosexual.* Grune and Stratton, New York.

Thompson, N.L., West, D. J., and Woodhouse, T. P. (1985). Socio-legal problems of male homosexuals in Britain. In *Sexual Victimisation,* (ed. D. J. West). Gower, Aldershot.

Trenchard, L. and Warren, H. (1984). *Something to tell you.* Gay Teenage Group, London.

Vassall, J. (1975). *Vassall: The autobiography of a spy.* Sidgwick & Jackson, London.

Winkelstein, W. *et al.* (1988). The San Francisco Men's Health Study: continued decline in HIV seroconversion rates among homosexual/bisexual men. *American Journal of Public Health,* **78**, 1472–4.

Whitam, F. L. (1983). Cultural invariable properties of male homosexuality. *Archives of sexual behaviour,* **12**, 207–26.

3
Male children and adolescents as victims:
a review of current knowledge
Bill Watkins and Arnon Bentovim

There are a number of prevalent myths about the sexual abuse of male children or adolescents. These include:

1. Boys are hardly ever abused sexually or that, when they are, the frequency of abuse is a tiny fraction of that of girls.
2. When boys are abused they are less psychologically affected than girls, both initially and in the long term.
3. If boys are abused the perpetrator is male and homosexual.

The definition issue

The normally accepted definition of sexual abuse is 'the involvement of dependent, developmentally immature children or adolescents in sexual activities they do not truly comprehend, and to which they are unable to give informed consent and that violate the sexual taboos of family roles' (Schechter and Roberge 1976). Researchers usually define the age difference between abuser and victim as needing to be of five years or more. However, with the recent concern about younger perpetrators, Johnson (1988, 1989) has advocated a two-year age difference, along with other criteria, to define sexual activities that are clearly abusive. Cantwell (1988) has advocated a definition for child abusers without specifying an age criterion for the perpetrator and focused entirely on the behavioural element. Thus, subjecting a child to oral-genital contact or penetration of the vagina or anus with fingers or objects would be regarded as abnormal.

There follows the issue of distinguishing between non-abusive and abusive contact with children. Sexualized attention has been proposed as a term to describe the interface between clearly abusive and appropriate interaction between adults and children (Haynes-Seman and Krugman 1989). Obvious abusive behaviour of a boy would be masturbating him, anal fingering or intercourse, but caressing and stroking a baby's buttocks persistently, or poking fingers in the child's mouth, which were arousing to the father, would be examples of what was considered as sexualized attention. Whether such contact is sexualized depends on the affective state of parent and child, and whether either has a sexual response to it. This clearly merges into the cuddling, vigorous bouncing, and rough-housing which is the normal approach to boys by fathers in particular.

A study by Chasnoff *et al.* (1986) of maternal-neonatal incest describes the impact of activities between mothers and infants that are clearly sexual, for example, sucking penises, and the subsequent powerful sexualizing effect this has on the baby boy's behaviour at a very young age. Yates (1982) also describes the eroticization of slightly older children by incestuous contact. Sroufe and Ward (1980) observe how mothers exerted control over their children, particularly their sons; these mothers had often been abused themselves and were therefore, quite unconsciously, reinacting their own experience.

Besides a few pioneering studies on selected, self-reporting groups of parents (e.g. Rosenfeld *et al.* 1986, 1987), we have little reliable information on the pattern of normal parent-child interaction in the community. Obviously, infants and toddlers are subject to close physical contact during care-taking functions. However, it is of interest that in the Rosenfeld *et al.* (1986) study more than 50 per cent of 8–10 year-old daughters were reported as touching, at some time, their mother's breasts and genitals, and more than 30 per cent of 8–10 year-old sons were reported as touching their mother's genitals and about 20 per cent their father's genitals. Current research (Smith and Grocke 1990), using direct interviews with children, will provide greater understanding of how the sexual knowledge of non-abused children differs from that of those who have been abused. As we have indicated, the prevalence and impact of sexualized contact in the general field of child rearing is only just beginning to be explored.

Another important issue is the effect modelling has on children (e.g. Seghorn *et al.* 1987; Smith and Israel 1987; Faller 1989 *a*) and whether, using an ecological model, there is sensitization of children through exposure to sexual activities at an inappropriately early stage of development, which subsequently triggers compulsive sexual responses (Yates 1982).

Prevalence of sexual abuse

Peters *et al.* (1986) have carried out a wide-ranging review of the prevalence of sexual abuse, comparing males to females, in samples of volunteers, college students, and subjects from the community at large. They concluded that there is considerable variation in the prevalence of child sexual abuse, ranging from 6 to 62 per cent for girls and 3 to 31 per cent for boys. Even the lower of these rates indicates that sexual abuse is far from an uncommon experience for either girls or boys, and the higher rates suggest a problem of epidemic proportions.

Although the majority of studies have shown that boys, like girls, are sexually victimized by men, recently some surprising findings have been reported. Fromuth and Burkhart (1989) studied two large samples of college men in the United States and reported a prevalence of 15 per cent in one area and 13 per cent in another. However, there are some unusual aspects to this study, including the fact that the majority of the perpetrators were reported as being female; 78 per cent of female perpetrators in one sample area and 72 per cent in the other. The investigators' definition included non-contact abuse, and a high proportion of the sample described only one 'abuse encounter'. Their findings illustrate how carefully attention has to be paid to the exact definitions of abuse used and the perpetrators involved. Fritz *et al.* (1981) also found that an excess of females (60 per cent) were reported to have abused a college sample of men, where molestation was defined as 'physical contact of an overtly sexual nature.'

Much of the epidemiological work discussed so far has been undertaken in the United States but British studies all indicate a high prevalence of sexual abuse in the United Kingdom, an increasing recognition of sexual abuse in the community, and an associated recognition of boys as victims (Mrazak *et al.* 1981; Baker and Duncan 1985, Creighton 1985; Bentovim *et al.* 1987; Markowe 1988).

Ratio of abuse: boys to girls

It has been suggested that girls are abused in a ratio to boys of 9:1 (Cupoli and Sewell 1988; Dubé and Hébert 1988), but other contemporary studies, including those in the community, indicate that sexual abuse of boys has been underestimated and that rates of abuse in girls are only a little higher than in boys (Keckley Market Research 1983; Badgley *et al.* 1984; Finkelhor 1984 *b*; Kercher and McShane 1984; Baker and Duncan 1985; Lewis 1985; Murphy 1985). Ratios reported from clinical samples show greater variation (Adams-Tucker 1984; Mian *et al.* 1986; Bentovim *et al.* 1987; Tong *et al.* 1987; Gale *et al.* 1988).

Prevalence in special populations

There are certain populations that require particular consideration because of their association with a higher prevalence of sexual abuse. These are:

(1) runaways;
(2) psychiatric in-patients;
(3) male prostitutes.

Runaway populations

It is estimated that between three-quarters of a million and 2 million children and adolescents in the United States run away from home each year (Shane 1989). McCormack *et al.* (1986), in a large series (144 adolescent runaways), found that 73 per cent of the girls and 38 per cent of the boys had been sexually abused. One assumption is that if girls run away it may be associated with sexual abuse, whereas boys run away for more diverse reasons. A note of caution should be sounded, however, as a broad definition of sexual abuse was used in that study. Stiffman (1989), using a contact-only definition of abuse, found a rate of only 10 per cent in a similarly large sample of runaways. Unfortunately no gender breakdown was given. Physical abuse was much more common (44 per cent).

Children in hospital

Children who are in hospital for psychiatric reasons are usually those children with the most severe symptoms, and the most worrying behavioural patterns, in the community. Just as sexual abuse and trauma are now emerging as major factors leading to psychiatric disturbance in adult women (Mullen *et al.* 1988), so a high incidence of sexual abuse is also being found in children who require psychiatric treatment (e.g. Emslie and Rosenfeld 1983; Husain and Chaple 1983; Kohan *et al.* 1987; Livingston 1987; Kolko *et al.* 1988; Singer *et al.* 1989). In a study of 54 subjects, 37.9 per cent of girls and 24 per cent of boys acknowledged a history of sexual abuse (Sansonnet-Hayden *et al.* 1987). Abused children had a greater severity of depressive symptoms, more hallucinations, made more suicide attempts, and were more likely to be referred for long-term, in-patient care. There was also a trend towards greater behavioural disturbance and longer hospital stays. Other studies of in-patient populations also show a higher incidence of abuse.

Child sex rings and prostitution

Certain sexual behaviours, defined as normal at one period and place in history, may be regarded as totally unacceptable at another (De Mause 1974; see also Chapter 7).

The secrecy surrounding sex rings has veiled our awareness of their extent (Christopherson 1990). It is only very recently that any information has become available in either America or the United Kingdom

(Burgess 1984 b; Wild and Wynne 1986; Finkelhor 1988; Dawson 1989).

Burgess et al. (1984 a) have defined a sex ring as 'being composed of an adult perpetrator (or perpetrators) simultaneously involved with several children who are aware of each other's participation in sexual activity'. Three types of rings are described: solo rings with only one adult; transitional rings focused on pornography; and syndicated rings where children are actively recruited on a commercial basis by a well-structured organization.

Groups advocating child sex and those lobbying for the liberalization of laws concerning the sexual exploitation of children have arisen since the 1960s. De Young (1988) has described members' self-justification of such activities by denial of injury, claims of victim culpability, condemnation of those who criticize or censure such organizations, and claims that they free children from the repressive bonds of society.

At times the boundaries of what constitutes a sex ring become blurred with those of large, chaotic families (Dawson 1989). Nevertheless, some have involved ritualized and satanic practices, including those of a very sadistic nature. Perpetrators have either enlisted children themselves or used child ringleaders to bring in new members. Offenders often target passive, quiet, lonely, troubled children in 'out of doors' places (Budin and Johnson 1989). Christopherson (1990) has described the impact on communities of discovering such rings, as well as the special staffing needs, costs, interviewing and therapeutic challenges that these events evoke.

Related issues

There is widespread anecdotal reporting of sexual activities within all-male institutions such as residential schools, but few studies to indicate the prevalence of such abuse. Abuse may be perpetrated by pupils within the school or be initiated by teachers. In a recent study it was reported that more than a quarter of perpetrators – the largest single group – were professionals of one kind or another (Faller 1989 b). Suppression of such events is common, as illustrated by a Canadian inquiry into abuse of boys in a orphanage in Newfoundland. Fourteen years after the initial investigation by police of sexual abuse, nine Christian Brothers and former Brothers have been charged (*Globe & Mail* 1989).

Under reporting

It is clear from community studies previously cited that under-reporting of sexual abuse is consistent and universal. It is probable that boys are even less likely than girls to report sexual assaults or have such assaults reported. Extrapolation of numbers of cases reported to professionals (Mrazek et al. 1981) leads to a prevalence figure for abuse of 0.3 per cent for the population (Peake 1990). This is in contrast to the figure of 8 per cent for boys revealed by Baker and Duncan (1985).

Some explanations for this discrepancy are as follows:

1. Children who are abused from early infancy through to the preschool period usually cannot perceive the abusive nature of their contact, particularly if the abuse has been preceded by extensive affectionate parenting, nurturing, and grooming, which subtly leads up to sexually abusive contact. It may well be that it is only through the alertness of those responsible for children that sexualized patterns of behaviour, or unusual sexual preoccupations, can be detected.

2. Even if the child does have some partial perception that what is occurring is abusive, he or she may not have sufficient mastery of language to describe what is happening. Various aids, such as drawings or anatomically correct dolls, have been used to assist where evidence is equivocal (Yates and Terr 1988 a, b).

3. Even if a child is capable of recognizing abuse and has the language skills to describe it, he or she may be inhibited by the consequences of disclosing, such as family disruption, punishment, or removal from home. Usually such fears are linked to actual or implied threats not to tell. Fear of consequences prevented all but one man in Dimock's (1988) sample of 25 men from telling anyone at the time their abuse was occurring.

4. Any child may be confused over the pleasurable aspects of interpersonal level, and thus be reluctant to disclose the abuse, particularly if it has been associated with affection of some kind. In boys, an involuntary erection may contribute to their sense of confusion, in contrast to girls where such obvious signs of physiological arousal are less obvious.

Under-reporting in boys

Most of the evidence suggests extrafamilial abuse is more common than incest (Finkelhor 1984 c; Rogers and Terry 1984; Baker and Duncan 1985; Tong et al. 1987; Vander Mey 1988; Faller 1989 b), but is divided over whether boys are more prone to abuse by strangers (Ellerstein and Canavan 1980; De Jong et al. 1982; Rimsza and Niggemann 1982; Baker and Duncan 1985; Spencer and Dunklee 1986; Bentovim et al. 1987; Reinhart 1987; Tong et al. 1987; Dubé and Hébert

1988; Friedrich 1988; Faller 1989 b). It is thought the police receive the most reports of extrafamilial abuse because it is more readily viewed as a crime (Finkelhor 1984).

A partial explanation for the greater extrafamilial incidence of abuse is the fact that older boys are less often supervised in the community (Budin and Johnson 1989) and are therefore more vulnerable. There may even be some myths about boys not needing protection (Pierce and Pierce 1985) as they are judged in terms of 'toughness', in contrast to the 'vulnerability' of girls. Similarly, there is possibly a tendency to discipline rather than help boy victims. In a study of the attitudes of college undergraduates, male victims who behaved in an encouraging manner were more likely to be 'penalized' by students of both sexes; male victims were penalized more by male students (Broussard and Wagner 1988).

Boys are more likely than girls to externalize their abusive and traumatic experiences (Friedrich 1988; Peake 1990). Sebold (1987) was one of the first to draw attention to the need to become aware of differential indicators of abuse for boys. Exhibitionism and sexual offending in pre-adolescent or adolescent boys are frequently overlooked as indicators that the boys may have been abused.

Factors influencing apparent prevalence in boys

Masculinity dynamics Boys are usually socialized into an ethos where self-reliance, independence, and sexual prowess are valued, while vulnerability and homosexuality are denigrated. Boys who show independence, initiative, strength, courage, agility, and fearlessness are admired. Disapproval is likely for crying or showing signs of dependency and helplessness. It is easy to see how the encouragement of these trends could lead to excessive aggressiveness, the abuse of power, and the denial of feelings, irrespective of innate biological factors (Nasjleti 1980; Finkelhor 1986; Peake 1990).

Fear of homosexuality Any child is likely to experience self-doubt, embarrassment, and shame because of their abuse. It is a common clinical experience for boys to feel that because they responded, it must mean that whoever victimized them knew they would react and had therefore picked them out because of some 'sign' of homosexuality. Sexual victimization may well have the effect of reinforcing such preoccupations and fears in regard of adolescent sexual identity, leading to repression or deletion of the experience, with a failure to report.

Even after disclosure, adolescent boys may be very reluctant to talk about their abuse in therapy, whether this be individual or group and irrespective of the therapist's gender (Nasjleti 1980). Such reluctance to talk may be associated with intense fears of homosexuality and suggests that they partially explain the failure of adolescents to report their abuse. Sebold (1987) believes homophobic attitudes in an adolescent male should be considered as a significant indicator of possible sexual abuse.

Many child-protection and clinical studies report that boys are abused at a younger mean age than girls (Ellerstein and Canavan 1980; De Jong et al. 1982; Rimsza and Niggeman 1982; Finkelhor 1984; Rogers and Terry 1984; Pierce and Pierce 1985; Bentovim et al. 1987; Singer 1989). However, other studies of community samples (Baker and Duncan 1985), students (Finkelhor 1979) and incarcerated sex offenders (Mohr et al. 1964; Gebhard et al. 1965; Frisbie 1969) have reported that the age at onset of abuse for boys was higher than for girls.

Intrafamilial abuse

All forms of intrafamilial abuse are under-reported because of the greater secrecy and the intense fears of the consequences that surround the abuse. The available evidence points to the increased possibility of boys being abused in conjunction with their sisters, rather than in isolation (Finkelhor 1984; Pierce and Pierce 1985; Bentovim et al. 1987; Vander Mey 1988; Faller 1989b), so it is not surprising that Reinhart (1987) reports a trend towards the abuse of boys being more often disclosed by a third party. Detailed interviews are needed of all the children in a family, including boys, when a girl reports intrafamilial abuse.

Not all reported clinical samples concur. Bentovim et al. (1987) had very few cases referred where abuse was perpetrated by anyone outside the family context. In the study by Friedrich et al. (1988) of 3 – to 8–year-old-boys, 95 per cent of the abuse was intrafamilial. Similarly, Hobbs and Wynne (1987) found a high proportion of boys abused within the family circle.

Abuse by women

Abuse by women, particularly mothers, has been an especially difficult issue for the community to contemplate. Even professionals lack an awareness that women can be abusers. The rate of abuse by women either jointly, in poly-incestuous activities, or alone is between 5–15 per cent of cases coming to professional note (Bentovim et al. 1987; Faller 1989 b). In Dimock's (1988) series of 25 abused men, 28 per cent were abused by women.

Sons were abused in 43 per cent of McCarty's (1986) series of mother-child incest, which Fehrenbach and Monastersky (1988) reported that 40 per cent of the children abused by female adolescent perpetrators were male. Finally, Johnson (1989) reports that in her small series of female perpetrators, boys were the victims in a ratio of two to every one girl. While Faller (1989 b) found boys 10 times more likely than girls to be victimized by a woman alone, Russell and Finkelhor's (1984) conclusion was that overall *more girls* than boys are victimized by female perpetrators.

McCarty (1986) noted that women are viewed as sexually harmless to children. Obviously abuse covers a much wider range of acts besides penile-vaginal intercourse. Sexual stereotyping may be responsible for under-reporting of abuse, because of a fear of disbelief, Krug (1989) and Singer (1989) both finding only one in their respective series of 8 and 12 men who reported the abuse by their mothers or other female relatives.

There is evidence that society tends to employ a double standard in which attitudes towards male socialization presents the redefinition of female-male abuse as simply a normative sexualization experience (Dimock 1988), or of no consequence because the abuse was frequently regarded positively (Fromouth and Burkhart 1989), or rated by the men retrospectively as having no effect (Baker and Duncan 1985).

Another stereotype is that, culturally, women are permitted a much freer range of sexual contact with their children than are men. Women usually bath, change, and dress their children. Society accepts mothers taking their daughters to bed (Banning 1989).

Because the sexual act generally requires the male taking the initiative or being 'active', confusion may exist in mother-son incest as to whether the abuse was perpetrated by the boy against his mother, or mother against the son. In fact, the large discrepancy in terms of sexual knowledge and sexual awareness between the mother and son makes the mother the abuser, despite the son sometimes feeling himself to be the active partner or even the initiator of sexual activities.

The issue of the perpetrator's was originally so powerfully enshrined in certain beliefs that it was maintained that any woman who could abuse her son must by psychotic. As recently as 1979, Rosenfeld claimed that 'in mother-son incest one or both parties is usually psychotic'. This clearly is not so and current clinical practice negates such views. Krug (1989) found no psychotic mothers in his series of cases of mother-son abuse, as judged by reports from the sons. In her study, McCarty (1986) found only two of the eight women who molested only male children had documented emotional disturbances. Our own clinical experience would indicate that the majority of women who are sexually abusing their sons may well have had a serious sexually abusive experience themselves, and indeed may well be continuing to be abused; the abuse therefore is occurring in the context of a poly-incestuous situation rather than frank psychiatric disorder.

Abuse of children by children and adolescents

Recent research has begun to focus on the need to recognize both the significance and the effects of sibling and cousin incest, of child-child and adolescent-child sexual abuse (e.g. Chasnoff *et al.* 1986; Smith and Israel 1987; Friedrich *et al.* 1988; Johnson 1988, 1989; De Jong 1989). These reports, which involve over 150 examples, show there is overlap between these various forms of abuse. They challenge further acceptance of sibling incest, in which there has been little interest, because it has not been perceived as being harmful (De Jong 1989). Despite the lack of documentation, sibling incest is thought to be the commonest form of incest (De Jong 1989). In one centre, child perpetrators are presenting at the rate of 3–4 each week (Cantwell 1988). Preceding sexual abuse is common amongst the male child/adolescent perpetrators whilst the only report on female-child perpetrators found it to be universal (Johnson 1989).

From the perspective of this chapter the key questions are:

1. How young can a child be and still be regarded as a perpetrators?
2. Is there a greater likelihood that female adolescent perpetrators will abuse boys?
3. Is there a greater likelihood that male adolescent perpetrators will abuse boys?

There is almost certainly insufficient information on which to base conclusions, although data from a number of studies are presented.

Depending on the laws in different countries there usually is no legal inclusion of children below a particular age, often 10 years, who abuse, so information from police or welfare sources is lacking. The youngest perpetrator found varies, as expected, with the series described – 9 years old by De Jong (1989) and Smith and Israel (1987), 6 years by Cantwell (1988) and 4 years in both Johnson's samples (1988, 1989). Chasnoff *et al.* (1986) describe a boy whose abuse stopped by the age of 9 months, who, at 25 months, was demonstrating 'sexual aggressiveness' towards other children. Is there a point at which 'sexualization' or 'acting out' is relabelled as 'abuse'?

As mentioned in the section on female abusers, 13 girl perpetrators in Johnson's (1989) series molested two boys for every one girl. From the data presented it is not possible, in Johnson's (1988) parallel series of boy perpetrators, to calculate a comparable ratio. Nevertheless, of the 47 boy perpetrators, 23 cases involved sibling incest, in which 52 per cent abused younger brothers. In the Smith and Israel (1987) study there were 5 female and 20 male perpetrators, including both children and adolescents. Their victims were overwhelmingly female – 89 per cent. Cantwell's (1988) interest was on cycles of abuse between adolescents/children and other children. It is difficult to unravel the figures, but many boys were involved as victim-perpetrators. The series of De Jong (1989) is also not comparable as the perpetrators, presumably all males, range up to 40 years of age, with an adolescent mean. Victims are again overwhelmingly girls, and while cousins are twice as likely to abuse a boy relative as brothers are, this does not reach statistical significance.

The question can, however, be approached from the perspective of offender reports. Evidence is accumulating to indicate that a majority of offenders begin their 'careers' in adolescence (Longo and Groth 1983). In another study of 561 adult sexual offenders, 59 per cent reported the onset of their paraphilic behaviour during adolescence (Abel et al. 1987). The uncertainty is over how typical or generalizable are the recent, rapidly rising number of studies on adolescent perpetrators (e.g. Shoor et al. 1966; Lewis et al. 1979; Fehrenbach et al. 1986; Becker 1988; Kavoussi et al. 1988). Shoor et al. found that 45 per cent of their Juvenile Probation Department sample ($n = 80$) had abused boys aged 10 years or less. These results are in keeping with those from an adult community-based, specialized, sex-offender treatment programme (Conte et al. 1989). Of the 20 men, selected to maximise the amount of information available, 35 per cent had abused boys. It is noteworthy that Budin and Johnson (1989) found that an almost identical number of their 72 incarcerated offenders (37.5 per cent) had abused boys. Lewis et al. found, in a group of 17 drawn from a secure unit for violent juvenile offenders, that 12 per cent had abused males. Fehrenbach et al. reported that of 297 adolescent perpetrators, 28 per cent abused males, mostly boys, nearly all of whom had had sexual offences committed against them, and the frequency of boys abused rose to be almost equal to that of girls for children aged 6 years or less. Becker reported that of 27 adolescent perpetrators with their own history of sexual abuse, 55 per cent abused boys. Finally, Kavoussi et al. found, in an out-patient sample of 37 adolescent offenders, that 38 per cent had abused boys aged 11 years or less.

For those who deal with adolescent perpetrators, it seems reasonable to conclude that the chances he will have abused a boy are fairly high. On the face of it these figures also bolster the case for an apparent under-reporting of boy victims. Some caution is required, though, because in the series of Conte et al. (1989) it is possible to calculate the total proportions of the victims – the 20 men abused 146 children, of which 34 were boys, which represents 23 not 35 per cent of the total. Pierce and Pierce (1985) think there may be a bias towards imprisoning abusers of boys, which would distort the above findings, but the actual number of cases proceeding to prosecution, whether boys or girls, was very low. Obviously these factors are important and they only emphasize how difficult it is to interpret the data.

Father-son abuse

In contrast to the number of publications about mother-son abuse, there is an astonishing lack of reports about father-son abuse, even though fathers are cited as the most frequent abusers of boys, including sons (Pierce and Pierce 1985; Spencer and Dunklee 1986; Vander Mey 1988; Hobbs and Wynne 1987; Reinhart 1987; Faller 1989 b). Stepfathers, as found by these investigators, tend to be the next most frequently cited perpetrators. Yet there are only sporadic case reports (Raybin 1969; Steele and Alexander 1981; Langsley et al. 1986), or small series of five to six cases (Dixon et al. 1978; Justice and Justice 1979) which have started to look at father-son abuse in any detail. The Justices (1979) explain this striking denial in terms of 'two moral codes: the one against incest and the one that has previously existed against homosexuality'. There are a few exceptions to the clinical finding that fathers/stepfathers are the most frequent abusers of boys (e.g. Ellerstein and Canavan 1980; De Jong et al. 1982), whose samples appear to be strongly slanted towards abuse by strangers. Johnson's (1988) report on boy perpetrators found that seven (30 per cent) had been abused by their fathers. The assertion by Langsley et al. (1968) that 'father-son incest is the least common [sexually abusive] combination' is unsupported.

There is some suggestion of there being an association between the physical and sexual abuse of boys (Finkelhor 1984; Spencer and Dunklee 1986; Sansonnett-Hayden 1987; Kolko et al. 1988; Cavaiola and Schiff 1989), particularly with father-son incest (Dixon et al. 1978) which, if true, has practical implications concerning awareness and detection. In this regard it is important to maintain a distinction between the kind of ongoing repetitive physical abuse, which occurs within the family, and the kind of force that is used during abductions. The latter is correlated with older boys, abuse by strangers, and oral/ perianal trauma (Ellerstein and Canavan 1980; De Jong et al. 1982; Rimsza and Niggemann 1982; Spencer and Dunklee 1986).

Pierce (1987), in her survey of the literature, could find only 52 instances of reports of father/stepfather sexual abuse. It is doubtful whether a case history (Halpern 1987) should be added to this number, as it involved the late adoption of a 12–year-old boy by a homosexual couple, where the abuser had an extensive paedophile history. In the main, paedophiles indicate a preference for boys (Righton 1981), and Finkelhor (1984) concluded that victimized boys are more likely than girls to come from impoverished and single-parent families, so it may well be that boys are more at risk from older paedophiles targeting single-parent families with male children. It must be recognized that a proportion of all abusers choose their families, their jobs, and friends solely with a view to gaining access to children (Peake 1990).

The nature of the abuse

Clinical reports, some without control groups, are unanimous in finding that boys are more likely than girls to be subjected to anal abuse (Ellerstein and Canavan 1980; De Jong et al. 1982; Rimsza and Niggemann 1982; Rogers and Terry 1984; Spencer and Dunklee 1986; Bentovim et al. 1987; Reinhart 1987; Cupoli and Sewell 1988; Hobbs and Wynne 1989). Hobbs and Wynne point out that the type of abuse varies by age. Girls are most likely to be anally abused when young, with a cross-over to vaginal abuse around the age of 10 years. If fingeranal penetration and object-anal penetration are included, then the percentages are even higher.

Despite popular belief to the contrary, physical findings are commonly present after anal abuse (Ellerstein and Canavan 1980; Spencer and Dunklee 1986; Reinhart 1987). These findings derive from paediatric settings, which may contain the more serious end of the range.

What constitutes a significant physical finding has recently been intensely debated. The Cleveland Inquiry in the United Kingdom (Butler-Sloss 1988) has, in particular, focused controversy over the significance of anal findings, including reflex anal dilatation (pp. 186–93). The report concluded 'we are satisfied from the evidence that the consensus is that the sign of anal dilatation is abnormal and suspicious and requires further investigation. It is not in itself evidence of anal abuse.' The latter point was made a recommendation (6. c, p. 247), along with the need to establish 'a consistent vocabulary to describe physical signs which may be associated with child sexual abuse' (6. a, p. 247).

The effects of sexual abuse

Just as there is an under-reporting of the incidence and prevalence of sexual abuse in boys, so there is a relative paucity of information about the effects of sexual abuse on boys, as distinct from the sexual abuse of girls. The question of whether there might be differential initial effects has yet to be seriously researched and we are left with scattered impressions from clinical settings, often involving small clinical samples.

General responses

Freidrich (1988) has proposed a framework which helps to conceptualize the information that is available. He has put forward the view that it is essential to use a developmental paradigm. The effect will depend on the cognitive level at which each is functioning. The younger child will be far more concrete about his/her abuse, perceiving it, for example, as abuse only if there is pain, whereas the older child will be able to conceptualize it in more abstract terms. The immaturity of the younger child increases the likelihood of regressive responses, while the older child may resort to running away or drug-taking. In addition, the family of the younger victim may respond quite differently from the family of an older victim.

A second paradigm which Friedrich (1988) presents as needing consideration is how actively does adaptation occur in response to abuse? Obviously what is being adapted to needs to be taken into account. Both Rogers and Terry (1984) and Friedrich et al. (1988) have suggested that greater initial symptomatology is associated with greater severity of abuse. Wyatt and Powell (1988) have been more specific and reach the general conclusions that the most negative consequences for children are associated with abuse by fathers, genital contact, and the use of force, the latter two being aspects of 'severity'. In similar vein, but also without gender analysis, Sirles et al. (1989) report that the presence of a Axis I, DSM-III (Diagnostic and Statistical Manual; American Psychiatric Association 1983) diagnosis is related to older victims, a closer relationship of the offender to the child, greater frequency, and longer duration of abuse.

Prerequisites for research into comparative outcomes between girls and boys are that groups should be matched in terms of demography, type of abuse, and abusers.

The extraordinary circumstances of purported abuse in a church-related, day-care centre, involving five boys and five girls aged 2 to 6 years, provided an opportunity to compare children. Factors including the timing of the abuse, place of abuse, kind of abuse, and abuser (Kiser et al. 1988). Various measures, including the Minnesota Child Development Inventory (Ireton and Thwing 1974) and the Child Behaviour Checklist (Achenbach and Edelbrook 1983), were used, and while initially the boys presented more clinically significant symptoms than did the girls, preliminary follow-up suggested girls were more symptomatic one year later.

With the increasing attention to post-traumatic states in children (Kiser et al. 1988; McLeer et al. 1988) there has been a concurrent interest in the coping strategies, competence, and resilience of abused children (Steele 1986; Mrazek and Mrazek 1987). Mrazek and Mrazek

Male children and adolescents as victims

propose a broad range of items that can be used to assess such factors. They are:

(1) rapid responsivity to danger;
(2) precocious maturity;
(3) dissociation of affect;
(4) information-seeking;
(5) formation and utilization of relationships for survival;
(6) positive projective anticipation;
(7) decisive risk-taking;
(8) the conviction of being loved;
(9) idealization of an aggressor's competence;
(10) cognitive restructuring of painful experiences;
(11) altruism;
(12) optimism and hope.

These items make an interesting counterpoint to the post-traumatic stress responses that are described in three basic areas: 're-experiencing' (e.g. flashbacks, re-enactments, sexualized talking and playing); and 'hyperarousal', (including poor sleep, irritability, and generally hostile aggressive tone). Finkelhor and Browne (1986) have described the way that such responses merge into the widespread personality reactions that they have categorized under the headings of: traumatic sexualization, powerlessness, stigma, and betrayal. Clearly this issue has to be broadened to look at the ecosystem of the family which can influence the individuals adaptation to the abuse. Bentovim and Kinston (1991) have described the way in which traumatic events can be deleted, with the creation of a hole 'in the mind of the individual', or within the familial or social context, which results in actions being substituted for thinking and working through. These responses may very well differ in boys and girls, because of differing approaches to, attitudes towards, and expectations of boys and girls in relationship to stress.

Specific responses

Rogers and Terry (1984) describe behavioural responses that are more or less unique to male victims, and that appear to be directly related to the homo-erotic implications of the sexual contact and differential cultural expectations of male behaviour. Specifically the common reactions recorded in boy victims were:

(1) confusion/anxiety over sexual identity;

(2) inappropriate attempts to reassert masculinity;

(3) recapitulation of the victimizing experience.

Confusion over sexual identity

The experience of a homosexual act contradicts the child's normal understanding of sexual relationships. In seeking an explanation for why he was selected, self-blame and guilt are common responses. A boy may attribute selection to a particular aspect of his physical appearance, his speech, his clothing, or any other personal characteristic that might be perceived as feminine and to have contributed to the assault. Through perceiving himself as effeminate, the boy blames himself for having attracted the abuser. If the boy does not physically resist, this may further challenge the boy's view of himself as masculine. He may be sexually aroused which creates a conflict in the child's sense of sexual identity and he may define himself as homosexual.

In a highly selected sample of adolescent in-patients, Sansonnett-Hayden *et al.* (1987) describe histories of cross-dressing in five out of their series of six sexually abused boys. It is not possible to correlate this finding with the identity of the perpetrator, which is unfortunate as one might expect greater confusion over sexual identity in father-son incest. There are some case reports hinting this may be so, for example, two of the four adolescents in the series of Dixon *et al.* (1978) are reported as having concerns about their sexuality. Homosexual anxieties are an aspect of the detailed father-son case history reported by Langsley *et al.* (1968), while cases of intergenerational homosexual incest are described by Raybin (1969) and Mrazek and Kempe (1981).

More broadly, a distinction needs to be made between a fear of homosexuality and the development of a homosexual preference. An associated question is, does child sexual abuse contribute to the expression of a homosexual preference? Finkelhor (1984c) writes that it is a traditional mythology that molestation leads to homosexuality. Rogers and Terry (1984) were impressed by the frequency with which parents of molested boys anxiously ask about the impact on subsequent sexual development. Sebold (1987), who interviewed 22 therapists working with sexually abused boys, felt that homophobic concerns may be of value as an indicator of abuse.

Finkelhor's (1984c) study of college students reported that boys victimized by older men were over four times more likely to be currently engaged in homosexual activity than were non-victims, and this applied to nearly half the boys who had had such an experience. Similarly, Johnson and Shrier (1987), who reported a significantly greater likelihood that boys molested by males would identity themselves as homosexual in contrast to those molested by females. Further, the adolescents themselves often linked their homosexuality to their experiences of sexual victimization. We do not know whether the relationship to the abuser, or specific type or duration of abuse, influences the effect on sexual identity.

Fromuth and Burkhart (1989) looked closely at sexual adjustment and behaviour in their two large samples of college men, and noted that sexually abused men were no more likely to report a homosexual experience occurring after the age of 12 years than were non-abused men. But, as they commented, the lack of relationship might be attributable to the majority of the perpetrators being female.

In clinical reports of adults (Bruckner and Johnson 1987; Dimock 1988; Krug 1989; Singer 1989), there are concerns about sexual preference in a significant minority of each group.

Two further points need to be made in relation to this topic. Firstly, it is clear that in general homosexuals do not report inappropriate sexual experiences in childhood, and secondly, only a tiny minority of homosexuals have any sexual interest in children (Newton, 1978. Finkelhor 1984). It is dangerous to equate a homosexual abusive act with an assumption that the perpetrator is homosexually orientated. As we have already described, sexual interest in children often encompasses children of either sex and involves men with a polymorphous sexual orientation that may extent to adults as well as children.

Inappropriate attempts to reassert masculinity

The emergence of inappropriate attempts to reassert masculinity is, in the opinion of Rogers and Terry (1984), the most common behavioural reaction of boy victims. Pre-abuse passivity or unassertiveness is followed by post-abuse aggression, manifested in fighting, destructiveness, marked disobedience, and a generally hostile or confrontational attitude. Such feelings and behaviour can be linked to the 'traumagenic dynamics' described by Finkelhor and Brown (1986). It is apparent from examination of these concepts that the psychological impact of each can be a potentially maladaptive, opposite reaction. One response to traumatic sexualization can be counter-assertion of masculinity, which may extend to aggressive sexual behaviour. The counter-response to powerlessness can also be aggression or excessive controlling behaviour. The response to the experience of betrayal is the development of powerful mistrust towards others. Stigmatization can be dealt with by stigmatizing others. The presence of these externalized coping strategies may contribute to the likelihood that an abused boy will become a perpetrator.

Boys are thought to demonstrate more externalizing responses to sexual abuse than girls. As Summit (1983) states: 'the male victim of sexual abuse is more likely to turn his rage outward in aggressive and antisocial behaviour. He is even more intolerant of his helplessness than the female victim ... child molestation and rape seem to be part of the legacy of rage endowed in the sexually abused boy.' However, the existing evidence on this issue is inconclusive.

The literature on classification of child psychopathology suggests two broad factors that subsume the majority of behaviour disorders in children; internalizing and externalizing (Friedrich 1988). These concepts underpin the well-validated Child Behaviour Checklist (CBCL) (Achenbach and Edelbrock 1983), which has been usefully deployed to show the effects on, and clinical course of, sexually abused children (Tong et al. 1987; Friedrich 1988; Friedrich et al. 1988; Kiser et al. 1988; McLeer et al. 1988; Stiffman 1989). From these some tentative conclusions are suggested. Sexually abused children are:

(1) significantly more likely to score in the clinical range of the CBCL than are community controls, particularly if suffering from post-traumatic stress disorder (see below) (McLeer et al. 1988);

(2) likely to have total scores in the same range as other psychiatrically referred but non-abused children;

(3) likely to have significant elevations of both internalizing and externalizing behaviours;

(4) in the case of young abused boys, reliably discriminated from conduct-disordered boys by the presence of sex problems and a lower degree of aggression;

(5) unlikely to show clear gender differences regarding either internalizing or externalizing factors (with the exception of severely abused girls, aged 6 years or less, who show more internalizing than abused boys of similar age (Friedrich 1988);

(6) strongly influenced by family variables, whether showing internalized or externalized behaviour problems (Friedrich 1988).

One final perspective on this question of externalizing versus internalizing can be obtained from the Los Angeles Epidemiologic Catchment Area study, involving a randomly selected community sample (Stein et al. 1988). Gender comparisons revealed that sexually abused men had 'acted out' lifetime and current psychiatric diagnoses such as drug abuse or dependence. Abused women, in contrast, had a higher prevalence of all lifetime disorders except antisocial personality, and a higher current prevalence of any disorder, major depression, and anxiety. Further research along these lines is clearly warranted.

Recapitulation of the victimizing experience

Although less common than confusion over sexual identity or compensatory aggression, there is a tendency among boy victims to recapitulate their own victimization, only this time with themselves in the role of perpetrator and someone else the victim (Rogers and Terry 1984; Cantwell 1988). Modes of inducing compliance, specific sexual acts, even age differences appear to be patterned after the original incident. One mechanism which would appear to facilitate the transition from victim to victimizer is 'identification with the aggressor'. Obviously mechanisms besides 'identification with the aggressor' are operating in some situations, for example, coercion and modelling (Friedrich 1988). In therapeutic work with adults who have abused children, one of the most powerful turning points we have experienced is when the man becomes aware that he has re-created his own abuse with other children.

The work of Conte and Schuerman (1988) supports the assertion that victimizing others is uncommon, subject to the reservation that there is no gender breakdown on the symptom checklist. While sexually victimizing others ranks thirty-first (2 per cent) out of 38 symptoms, it still ranked ahead of suicide attempts (thirty-second) and other forms of self-harm (thirty-third). Sansonnett-Hayden *et al.* (1987) report 3 out of 6 abused adolescent boys become perpetrators, while Friedrich *et al.* (1988) found 4 (13 per cent) out of 31 boy victims had become perpetrators by the age of 8 years. The most extraordinary report in this regard is that of Chasnoff *et al.* (1986) where of three baby boys, whose abuse stopped at 4, 9 and 18 months, respectively, two had begun before the age of 3 years to sexually molest other children. Recapitulation of the victimizing experience is a finding of grave concern and of great practical importance. Should it be replicated, then boys who have been sexually abused need to have included in their therapy programmes strategies to prevent them becoming perpetrators. A problem arises as to which boys are particularly vulnerable and need to be targeted for this kind of work.

Finkelhor (1986) has argued cogently about the dangers of a single-factor theory whereby victims become victimizers – that it is exaggerated, that it ignores sociological aspects, that it will strike terror into the hearts of victims, and that worse, it might become a self-fulfilling prophesy. Further, it is quite clear that not all abusers were themselves abused. We agree with and share these concerns, but also believe if the evidence continues to support such a hypothesis, then it cannot be ignored, particularly if it offers a valid preventative strategy.

There are a number of reports of group work with men who have been sexually abused that report on men's concern about their propensity to abuse children (Singer '89, Bruckner and Johnson '87, Dimock '88).

It is clear that those who work with perpetrators, particularly adolescent offenders, think the victim-abuser cycle is relevant (Becker 1988; Cantwell 1988; Faller 1989a; Freeman-Longo 1986; Ryan et al. 1987; Ryan 1989). Kaufman and Zigler (1987) have warned of the impact of experimental design on perceptions of the strength of associations. They cite a study of physical abuse and neglect where retrospective analysis indicated a 90 per cent rate of intergenerational transmission, while prospective analysis indicated a rate of only 18 per cent.

Researchers who have reported on series of child and adolescent perpetrators describe widely divergent rates of abuse in the backgrounds of those they have studied. Jones et al. (1981) found that none of their 24 offending adolescents reported of abuse. Pomeroy et al. (1981) found 1 out of 6 perpetrating boys and adolescents had been sexually abused, although the details of how they assessed the boys for sexual abuse were not given. Other earlier reports show that the question of prior sexual abuse was not even considered (Shoor et al. 1966; Lewis et al. 1979) which probably reflects the lack of recognition of the time.

Recent reports concerning boy and adolescent abusers give a different picture. Both Fehrenbach et al. (1986) and Becker (1988) found rates of 19 per cent prior sexual abuse in a combined total of 422 adolescent offenders. The most striking findings are those of Smith and Israel (1987), where 52 per cent of their sibling perpetrators' sample had previously been abused; and Johnson (1988), where 49 per cent of her male-child perpetrators had been previously abused. Longo (1982) reported that 47 per cent of the adolescent sex offenders on his treatment programme had been sexually abused.

The findings with sexual offenders against adults complement what is described with child abusers. Of 106 child molesters, 32 per cent reported some form of sexual trauma in their early development (Groth and Burgess 1979). The University of Michigan Interdisciplinary Project on Child Abuse and Neglect found, in those where information was available, that 27 per cent of the intrafamilial perpetrating fathers or stepfathers had been abused (Faller 1989a). Seghorn et al. (1987) studied the entire population of the Massachusetts Treatment Centre for Sexually Dangerous Persons and found that 57 per cent of the 54 child molesters had been victims of childhood sexual assaults (rapists had less than half this prevalence – 23 per cent). An even greater divergence was found by Pithers et al. (1988), where 56 per cent of 135 paedophiles and only 5 per cent of 64 rapists had histories of childhood sexual victimization. It has been argued by Freeman-Longo (1986) that abuse involving multiple abusers or repeated abuse of long duration is more influential. Russell and Finkelhor (1984) believe perpetrator outcome is similarly associated with more severe, more unusual, and more disturbing abuse.

In summary, current evidence supports the conclusion that the sexual abuse of boys in childhood is an important contributory but not a necessary factor in the development of a perpetrator. For girls, prior abuse may be a necessary perpetrator developmental factor. Any child who is referred because of concerns about sexually abusive behaviour towards other children should themselves be assessed for possible abuse.

Identification of perpetrator risk in childhood

Becker (1988) has proposed a broad contextual model to encompass sexually abusive adolescents, which includes interaction between individual, family, and social variables.

We need to improve our skills in the prediction and identification of those children who have become, or are at risk of becoming, perpetrators. One of the puzzles of child sexual abuse is why so few girls in contrast to boys become abusers after their own abuse. Finkelhor (1986) has challenged all theorists to explain, within their model, the 'male monopoly' on molestation and the fact that not all victims become victimizers. Finkelhor then outlines a variety of possible explanations, such as that women are socialized to be more sexually submissive, boys may have more childhood sexual experiences than girls, boys may be physiologically aroused more quickly and sexually conditioned more easily than girls, while girls, through the selective promotion of nurturing roles, may have more internal inhibitions to overcome.

Three studies using standardized measures have now confirmed that there is a relationship between inappropriate sexual behaviour and sexual abuse (Tufts' New England Medical Centre 1984; Friedrich et al. 1988; Gale et al. (1988) found that sexualization significantly discriminated sexually abused children from abused and non-abused children. The survey by Kohan et al. (1987) of 110 children's units, the top four behaviour items listed in association with sexual abuse were sexual in nature. Conte and Schuerman (1988) convincingly demonstrate, however, in their much larger sample that sexualization is by no means a universal response to abuse. Although we do not know whether the finding holds equally for girls and boys, only 7 per cent of the children were regarded as showing age-inappropriate sexual behaviour.

Aspects of sexualization and strong emotional or behavioural reactions to neutral inquiry about genital anatomy or 'exposure to differentiated sexual experiences' have been proposed as the primary features of the 'sexually abused child's disorder', over which there has been intense debate regarding inclusion in the American Psychiatric Association's Diagnostic and Statistical Manual (Corwin 1988). One of the main conceptual dilemmas, apart from its validity, was whether it should be a stand alone or subsumed within post-traumatic stress disorder (PTSD) (Corwin 1988). The key issue is whether or not sexualization should be regarded as a re-experiencing phenomenon (Kiser et al. 1988; McLeer et al. 1988).

Finkelhor (1988) has observed that not all sexually abused children develop a PTSD and that sexualized behaviour could be better accounted for by learning or conditioning theory. Wheeler and Berliner (1988) have also presented a view that the heterogenicity of effects of sexual abuse are best accounted for by classical and social learning theory. Through the offenders' modelling, instruction, direction, or differential reinforcement, plus the threat of or actual use of punishment, abused children acquire a repertoire of sexual behaviours. Further, these behaviours have been acquired before the child has the necessary emotional, cognitive, or social capabilities to regulate their own sexuality.

A question still remains over how any of these theoretical models can account for the driven, compulsive quality observed in the post-abuse sexual behaviour of some children. As Wheeler and Berliner note, autonomic arousal may have a direct facilitative effect on the acquisition of sexual behaviours. There is abundant evidence of differences between individuals in the reactivity of their autonomic system, which may be a factor in the outcome for any one particular child. What is striking about these children is their resistance to change in alternative environments and the ineffectiveness of sanctions and prohibitions.

The problem is well recognized by foster parents who are frequently at a loss over how to cope with such behaviour. The very young age at which this occurs is both distressing and peturbing. Dramatic case histories are provided by Yates (1982) and Friedrich (1988) involving children of 2, 6, and 8 years.

With regard to the question of whether sexualization after sexual abuse leads to similar behaviours in adulthood, it has long been noted that one possible outcome for abused girls is promiscuity (Finkelhor 1986). Krug (1989) reported that six out of eight of his abused male patients described having multiple concurrent sexual partners. Dimock (1988) goes further and considers sexual compulsiveness to be one of the common characteristics of abused men.

Table 3.1 *Perpetrator risk index following child sexual abuse*

Gender	Male (possibly linked to temperamental factors)
Abuser	Male
	Close relative
	Multiple perpetrators
Type of abuse	Repeated
	Long duration
	Greater severity
Age of child	Impact greater with young child (less than 8 years)
Effects	Anxious sexualization (linked to own CSA*)
	Externalization coping adaptation (linked when present to own physical abuse)
	Sexual identity confusion; identification with the aggressor.
Diagnosis	Conduct disorder
	Post-traumatic stress disorder
	When present, all link to:
	Attention deficit disorder
	Learning difficulty
	Increased likelihood of:
	Pan-immaturity; externalization
Treatment	Treatment for own CSA not sought/provided
	Treatment for early penetrating behaviour not sought/provided
Family	Intergeneration CSA history. Weak sexual boundaries (seen as seductiveness, extramarital liaison, exposure to pornography, denial of perturbing sexual behaviour in own children)
	Prohibition of sexuality
	Coercive beliefs and behaviours regarding sexuality
Social	Isolation from peers
	Socialization into 'male dominance' culture

*CSA + Childhood sexual abuse.

We believe it is possible to draw together the disparate strands regarding the identification of future perpetrator risk. Our model, which is drawn on that of Becker (1988), is shown in Table 3.1. The key factors are a combination of sexualization and externalization. Virtually every factor listed augments either one or both of these processes. We regard them as facilitating the development of the emotional congruence, while Finkelhor (1984) has postulated as one of the four preconditions to sexual abuse. At present such an index cannot possibly be regarded as validated as it is essentially based on our own experience and the clinical data we have discussed. Nevertheless, we think it has heuristic value in the selection of certain boys for treatment and for the development of relevant research strategies.

Sexualization and externalization can be conceptualized as threshold phenomena, hence their value in partially explaining why some, but not all, boys later become perpetrators. Even when present they are open to modification by a host of factors besides treatment. It would be premature to use such an index in a closed predictive way and it would be regrettable if any tone of inevitability was conveyed to the care-givers of these children. Instead our intention is to provide a framework whereby a rational basis can be given for explaining concern.

The link to physical abuse operates through the intensification of anger, which in turn increases the likelihood of externalization. Traditionally, society sanctions much freer expression of anger by boys than by girls. It is doubtful whether much of the literature on the effects of sexual abuse has made sufficient allowance for the confounding influence of physical abuse.

What this model fails to do is to contribute to an understanding of the developmental track pursued by those boys who have not been sexually abused, yet who have become abusers. Some of the more general factors could be equally applicable but it is doubtful that they are sufficient to explain a perpetrator outcome.

Long-term effects

Our discussion so far has focused on the initial, general responses reported in boys, as well as those regarded as more specific, such as confusion of sexual identity, attempts to reassert masculinity, and recapitulation behaviours.

The assessment of long-term effects raises a number of key questions:

1. Is there a demonstrable association between childhood sexual abuse and later psychological disorder which significantly exceeds that of non-abused males?

2. If there is, has disorder been continuously present or has onset occurred later on in life?

3. What proportion of sexually abused males have an associated disorder and does this proportion differ between men and women?

4 Does the pattern of disorder/difficulty seen differ between men and women?

Several studies are now appearing that allow at least a preliminary evaluation of some of these questions. As Briere and Runtz (1988) point out, an association in women between sexual abuse and later psychological and social dysfunction can no longer be dismissed with statements like 'research is inconclusive' (Henderson 1983). They also note the wide range of opinion, which extends from claims that abuse is almost inevitably destructive to those who see abuse as a potentially positive experience.

Long-term psychological effects

A major problem in retrospective epidemiological studies that include men is how effectively they identify subjects who have been anally abused (e.g. Finkelhor 1979; Fritz et al. 1981; Stein et al. 1988; Fromuth and Burkhart 1989). From the information available in these studies, few, if any, questions regarding abuse have covered this possibility, yet it seems reasonable to assume that anal abuse would be the kind of abuse adult men will be most reluctant to admit spontaneously. Such omissions will not only contribute to under-reporting, but may also bias efforts to analyse long-term associations.

Baker and Duncan (1985) showed that despite equal rates of contact sexual abuse, males reported themselves as being significantly less damaged by their abusive experience than did females. The researchers suggested that boys might more readily dissociate from the experience because it is not congruent with expected adult sex-role behaviour. Rogers and Terry (1984), however, believe it is the homosexual nature of the act, that leads to the most psychological conflict. Differences will occur depending on the population studied, whether contact or non-contact abuse is involved, the duration of the abuse, and whether the perpetrator is male or female.

When psychiatric disorders are looked at, two interesting findings emerge. Firstly, in a clinical study, Briere et al. (1988) looked specifically at the symptomatology in men who had been sexually abused as boys and who later presented to a crisis centre. What was striking was that the men had a range of disorders very similar to that to their abused female counterparts. Both were equally likely to have made previous suicide attempts and significantly more so than non-abused controls. This accords with Singer's (1989) clinical impression of men and McCormack et al.'s (1986) findings on abused adolescents.

Using the Trauma Symptom Checklist (TSC-33), Briere et al. (1988) found that both abused men and women manifested greater syptomatology in all instances (dissociation, anxiety, depression, anger, sleep disturbance) than their non-abused controls. By definition, a study of this nature will miss out those men and women whose response to abuse was to act out, rather than to experience the sort of symptoms that would lead them to a health centre.

The most rigorous information comes from a large-scale Los Angeles Epidemiological Catchment Area Study (Stein et al. 1988) which shows that sexually abused men had *higher* prevalence rates (on both the lifetime and 6–month prevalence of any psychiatric diagnosis,) than women. This excess is entirely accounted for by the greater frequency in men of substance abuse disorder and, at least on the lifetime prevalence figures, of antisocial personality disorder. Women had higher rates of anxiety and depressive disorders than men. There was no excess of schizophrenic disorders.

It is tempting to link these outcomes to the possible gender differences in the likelihood of externalizing or internalizing, which has been covered earlier in the section on the initial responses of children. On this note, Briere et al. (1988) found abused men to be the most angry group, followed by abused women and then the control groups.

Impact on adult sexual functioning

Adverse effects on adult sexual functioning are frequently described in sexually abused girls. Much less is known about men and the results are confusing. Johnson and Shrier (1987) found that a significantly greater proportion of abused adolescents described sexual dysfunction (inhibition of libido, premature ejaculation, erectile difficulties, and failure to ejaculate). Overall, Fromuth and Burkhart (1989) did not find such difficulties, although one of their two groups had a greater likelihood of premature ejaculation and the other group had erectile difficulties. Pierce (1987) cites three studies where sexually abused sons later marry and continue to have sexual problems. Stein et al. (1988) found a consistent trend in terms of lifetime prevalence whereby twice as many women as men reported fear of sex, lowered libido, and less sexual pleasure. The same questions covering a 6–month prevalence found no abused men describing fear of sex, lowered libido, and less sexual pleasure. An intriguing point raised by this last finding, is whether there are gender differences in the likelihood of sharing feelings. Men are generally not renowned for their ability to express their feelings, in fact the opposite holds true and they are particularly loathe to admit to feelings of sexual inadequacy or difficulty. The problem is to know whether a 'No' represents a genuine answer or is a reflection of a psychological state of denial. Stein *et al.* (1988) only focused on inhibition of sexuality and did not cover compulsive or disinhibited sexual behaviours. At a more global level, McCormack et al. (1986) found that sexually abused runaway girls were significantly more likely to have confused feelings about sex than non-abused girls. The same significant difference was not found with males. Finkelhor (1984) and found that abused men had lower sexual self-esteem than abused women, and both had lower self esteem than non-abused controls.

Substance abuse

Uncontrolled clinical studies are unanimous in reporting problems of substance abuse in abused males (Bruckner and Johnson 1987; Dimock 1988; Krug 1989; Singer 1989). Two recent papers on adolescents provide a preliminary insight into how sexual abuse might evolve into substance abuse. First, Cavaiola and Schiff (1989) found, in a residential treatment centre for chemically dependent adolescents, that sexually abused subjects (male and female) had begun use of either alcohol or drugs at a significantly younger age than their control groups drawn from the same treatment centre and local high schools. Second, Singer *et al.* (1989) addressed the question of temporal sequence and found that 77 per cent of their adolescent psychiatric in-patients had been sexually abused before or concurrently with their first drink or first use of drugs. Then they showed that severity of alcohol and drug misuse, as judged by number of times drunk or 'high' on drugs, was significantly associated with sexual abuse. The proportion of sexually abused adolescents regularly using cocaine and stimulants was greater. No gender analysis was undertaken in either of these studies and at present we must assume the results apply equally to male and females.

Self-esteem

Cavaiola and Schiff (1989) report that low self-esteem is an enduring sequel to abuse. All subscales of the Tennessee Self-concept Scale were scored significantly lower in abused adolescent runaways than in non-abused controls. Stiffman (1989) has not been able to replicate these self-esteem findings using a different inventory, even though the runaways showed significantly more behavioural problems and depression. It is difficult to explain such discrepant results, which include failure to find an association with substance abuse. They may be a reflection of the sample characteristics – the adolescents came from grossly dysfunctional family backgrounds. Paradoxically, with regard to lower self-esteem, anxiety, and depression, Fromuth and Burkhart (1989) obtained no association with sexually abused men. These findings are very much in keeping with the prediction by Haugaard and Emery (1989) that a researcher using a broadly defined victim group would conclude that the experience of sexual abuse is quite different for boys and girls, while a researcher using a narrow victim group would conclude that there are few differences.

Effects on future relationships

Reports from uncontrolled clinical studies are in agreement that severe difficulty is experienced in maintaining sustained and meaningful relationships (Bruckner and Johnson 1987; Dimock 1988; Krug 1989; Singer 1989). This results from a mistrust of others, fear of intimacy, making and breaking relationships abruptly, and lastly from re-creating abusive relationships that echo the shortcomings of the childhood relationship. All of these factors are likely to spill over into difficulties with sexual relationships in general.

In a controlled study of sexually abused adolescents, McCormack et al. (1986) found that males had more difficulty interacting with friends, with same-sex relationships and with opposite-friend relationships. Compared to non-abused controls, abused adolescents had a significantly greater fear of adult men.

Summary

There is now preliminary evidence which shows that there are significant, adverse, long-term effects for sexually abused males, particularly when such abuse involves physical contact by other males. While this is true of both anxiety disorders and depression, it is especially true of substance disorders.

Management and treatment

The process of recognizing, reporting, and diagnozing boys who have been sexually abused has many similarities to that of girls. There remain, however, some important clinical issues, such as who should interview boys, because it may well be felt that a male interviewer will recreate the traumatic experience through a generalization effect. However, although it may be advisable for women interviewers to be used in interviewing both girls and boys who have been abused, it is possible that some boys are better able to respond to male interviewers.

Because it is believed that boys are abused in conjunction with their sisters and because perpetrators are more willing to take responsibility for the abuse of girls rather than boys, it is very important, clinically, to review the experiences of all children in the family. There has been a widespread view within protective and treatment services that the abuse of boys is not so relevant, should be discounted and is not particularly harmful (Nasjleti 1980).

Development of treatment services in the Great Ormond Street child sexual abuse project

The project began in 1981, working initially with a series of older girls who had been rejected by their own families. They formed the first group (Furniss et al. 1988), along with a group that was run in parallel for two of the mothers. The development of the project had been very much influenced by the approach of Giaretto, who advocated and developed a guided, professional, self-help approach to working with sexually abused children and their families (Giaretto 1981). This approach offers groups to children and young people of different ages. It also offers groups for mothers, perpetrating parents, and couples, closely linked with counselling for individuals for families.

At Great Ormond Street a similar approach was adopted (Bentovim et al. 1988). Along with the exponential increase in referrals throughout the 1980s, there was a rapid increase in the number of boys of all ages seen. Overall, in our series of 411 children, contrasted to siblings, there were 77 per cent female and 23 per cent male victims, 43 per cent non-victim females and 57 per cent non-victim males. We have already commented on the differences in age of onset and the extensiveness of abuse found with the boys. A decision was made to have group work, of increasing length relating to the child's age and stage of development, with the aim of achieving homogeneous groups of children.

Early on in the work with children a decision was taken to work with abused adolescent boys separately from adolescent girls who were victims. Although Giaretto (1981) has put forward the possibility of adolescent boys and girls working together, it has been our experience that group work with boys aged between 12 and 14 years works best when the boys are on their own. Optimally, co-therapists need to be male and female, which recreates a parental model. Our experience indicates that while it is possible to meet with similarly aged girls for up to 15 or 20 sessions, with adolescent boys it is often not possible to go beyond about 12 meetings without the development of significant disruption and acting out. The goals of such work have been to provide a regular opportunity for boys to share their experiences, to clarify confused feelings and ideas about their abuse, and to educate them in ways of extracting themselves from risk situations.

The nature of the themes needing examination have been described by Furniss (1990) and Rogers and Terry (1984). The following have been important topics we have found worth pursuing:

1. How and why might an adult abuse a child? This helps the child construct a cognitive picture of how it is that men, in particular,

might develop sexual interests in children and therefore in themselves.

2. The next important theme to explore is who is responsible for child sexual abuse and to emphasize that it is always the adult. As we have already indicated, we believe it is only those boys who have been most severely abused who are coming to clinical attention and that it is these very boys who are most at risk of becoming, or have actually become, perpetrators. Thus an important decision has to be made about whether to separate boys who are also abusing from those who have only been offended against. Clinical experience would indicate that it can be extremely difficult for a boy who has been abused, perhaps by an older boy, to remain in a group with another young person who has begun to perpetrate in addition to being a victim.

These choices are complex and require careful consideration, if the group context itself is not to become abusive.

Aspects of the group work

Feelings aroused by sexual abuse A whole set of feelings aroused by sexual abuse need to be explored, including feelings towards perpetrators, towards the important adults in their lives, towards themselves over gender identity, and over being able to make friends. It is likely some fears will be projected on to the peer group, such as fears of being regarded as effeminate or a sissy. Future fears appear to be most strongly focused on becoming homosexual, on becoming an adult who will abuse other children, and on having major problems in heterosexual relationships. Being in a group permits discussion of these issues and increases awareness that others have the same fears and expectations.

Although boys in this age group may well be able to talk about themselves, it is often extremely helpful to have ways in which the boy can look outwards rather than inwards. Questionnaires, card sorts of different responses, viewing videos, are all ways that promote adolescents seeing their problems from an alternative perspective, without undue reliance on talking about feelings and responses, which threaten to overwhelm them.

Relating individual work with boys to work in the family context An essential component of the model developed in Great Ormond Street is that work with the family needs to be integrated through stages, particularly when abuse has been intrafamilial as opposed to extrafamilial. The stages of treatment are:

1. Work during the phase of disclosure. During this phase an essential first step is for the abuser within the family to begin the process of taking responsibility for the abuse. Professionals need to establish the degree of concern, warmth, and empathy from the mother, or whether instead the child is being subjected to a scapegoating and blaming. In many families where boys and girls are jointly abused there is a sense that there is more sympathy for the girls and somewhat less for the boys. In those families where boys have been abused by outside members of the community, the main problems for the parents are their struggle with enormous anger towards, and blaming of, the perpetrator, and with great frustration with workers in the community, whom they may feel are showing more sympathy towards the offender than the child who has been victimized. The sense of grievance in these situations is often very powerful. It is essential that such parents should have an opportunity of working together and sharing these feelings with other parents.

Parents must grapple with particularly difficult feelings when an older child, for example, a boy in a family, abuses a younger sibling, whether a boy or a girl. In these families there is tremendous confusion over expressing sympathy towards the child who has been victimized while at the same time feeling rage towards the other child who is the perpetrator. It is essential to help the parents keep a balance between their concern for the child who has been victimized and the needs of the older child, who so often has been previously victimized himself.

A breakthrough point for boys who have been abused, or boys who are abusing and their parents, is being able to get in touch with the pain and helplessness associated with the traumatic experience rather than deleting these experiences from consciousness or becoming preoccupied with blaming others. Furniss (1990) makes the important point that boys must take responsibility for abusing others before their own victimization can be addressed, otherwise there is a very real danger that the child/adolescent perpetrator will use their own abuse both to avoid taking responsibility and as a perpetual excuse.

2. Work during the period of separation. Separation needs to occur when the victim is not supported by the non-abusing parent within a family. If there is a high degree of scapegoating or blaming, then the boy needs to be cared for in a separate context, preferably a foster home. Those children who are in a transition from victim to victimizer may need placement in a therapeutic community setting, which can contain whilst still protecting other children and yet continue to work intensively with the boy who is beginning an abusive pattern.

It is much more preferable for the abusing parent to be the one living separately so that the boy who has been abused against can live with his mother. Even so, work often needs to focus on improving the parental relationship between mother and son, and to decrease the sense of distance between them, which may have been produced by the intense identity confusion of the boy. This confusion can lead to the boy pulling away from his mother just as much as his father. During this phase, group work is often helpful in conjunction with more individual counselling approaches. At the same time the abusing and non-abusing parent need to work with their peers so that there can be a strengthening of the non-abusing parent's autonomy, in place of the control previously exercised by an offending parent. The family modes of silencing weaker members also need to be broken down. Some family meetings are helpful during this phase but no permanent change can be expected through family meetings as the family rarely lives together.

3. Rehabilitation phase. In those families where group and individual work can be seen to be resulting in major changes of potentially abusive and traumatic responses, then some testing out of increasing contact can occur, with the eventual possibility of rehabilitation in mind. It is essential that the mother has a one-up position of authority over the care of the children in the family, so that effective rules can be created which guard against further abuse.

References

Abel, G. G., Becker, J. V., Mittelman, M., Cunningham-Rathier, J., Rouleau, J., and Murphy, W. (1987). Self reported sex crimes in nonincarcerated paraphilias. *Journal of Interpersonal Violence*, 2, 3–25.

Achenbach, T. M. and Edelbrock, C. (1983). The child behaviour checklist and revised child behaviour profile. Queen City Printers, New York.

Adams-Tucker, C. (1984). The unmet psychiatric needs of sexually abused youths: referrals from a child protection agency and clinical evaluations. *Journal of the American Academy of Child Psychiatry*, 23, 659–67.

American Psychiatric Association (1983). *Diagnostic and statistical manual of mental disorders*, (3rd edn.). APA, Washington DC.

Badgley, R. *et al.* (Committee on Sexual offences Against Children and Youth) (1984). *Sexual offences against children*, Vol. 1. Canadian Government Publishing Centre, Ottawa.

Baker, A. W. and Duncan, S. P. (1985). Child sexual abuse: a study of prevalence in Great Britain. *Child Abuse and Neglect*, 9, 457–67.

Banning, A. (1989). Mother-son incest: confronting a prejudice. *Child Abuse and Neglect*. 13, 563–70.

Becker, J. V. (1988). The effects of child sexual abuse on adolescent sexual offenders. In *Lasting effects of sexual abuse,* (ed. G. E. Wyatt and E. J. Powell). Sage, Beverley Hills.

Benedek, E. P. and Schetky, D. H. (1987). Problems in validating allegations of sexual abuse. Part I: factors affecting perception and recall of events. *Journal of the American Academy of Child and Adolescent Psychiatry,* **26**, 912–15.

Benedek, E. P. and Schetky, D. H. (1987). Problems in validating allegations of sexual abuse. Part 2: clinical evaluation. *Journal of the American Academy of Child and Adolescent Psychiatry,* **26**, 916–21.

Bentovim, A., Boston, P., and Van Elburg, A. (1987). Child sexual abuse – children and families referred to a treatment project and the effects of intervention. *British Medical Journal,* **295**, 1453–7.

Bentovim, A., Elton, A., Hildebrand, J., Tranter, M., and Vizard, E. (1988). *Child sexual abuse within the family: assessment and treatment.* Butterworth, London.

Bentovim, A., and Kinston, W. (1991). Focal family therapy. In *Handbook of Family Therapy,* (ed. A. Gurman and D. Knistern). Basic Books, New York.

Briere, J. and Runtz, M. (1988). Post sexual abuse trauma. In *Lasting effects of child sexual abuse,* (ed. E. G. Wyatt and E. J. Powell). Sage, Beverley Hills.

Briere, J. and Runtz, M. (1989). University males' sexual interest in children: predicting potential indices of 'paedophilia' in a non-forensic sample. *Child Abuse and Neglect,* **13**, 65–75.

Briere, J., Evans, D., Runtz, M., and Wall, T. (1988). Symptomatology in men who were abused as children: a comparison study. *American Journal of Orthopsychiatry,* **58**, 457–61.

Broussard, S. D. and Wagner, W. (1988). Child sexual abuse: who is to blame? *Child Abuse and Neglect,* **12**, 563–9.

Brucker, D. F. and Johnson, P. E. (1987). Treatment for adult male victims of childhood sexual abuse. *Social Casework,* **68**, 81–7.

Budin, L. E. and Johnson, C. F. (1989). Sex abuse prevention programmes: offenders' attitudes about their efficacy. *Child Abuse and Neglect,* **13**, 77–87.

Burgess, A. (ed.) (1984a). *Child pornography and sex rings.* Lexington Books, Lexington, MA.

Burgess, A. (1984b). Response patterns in children and adolescents exploited through sex rings and pornography. *American Journal of Psychiatry,* **14**, 656–62.

Butler-Sloss, E. (1988). *Report of the enquiry into child abuse in Cleveland 1987.* HMSO, London.

Cantwell, H. B. (1988). Child sexual abuse: very young perpetrators. *Child Abuse and Neglect,* **12**, 579–82.

Cavaiola, A. A. and Schiff, M. (1989). Self esteem in abused in abused chemically dependent adolescents. *Child Abuse and Neglect,* **13**, 327–34.

Chasnoff, M. D., Burns, W. J., Schnoll, S. H., Burns, K., Chisum, G., and Kyle-Spore, L. (1986). Maternal-neonatal incest. *American Journal of Orthopsychiatry,* **56**, 577–80.

Christopherson, J. (1990). Sex rings. In *Working with sexually abused boys: an introduction for practitioners*, (ed. A. Hollows and H. Armstrong). TAGOSAC, London.

Conte, J. R. and Schuerman, J. R. (1988). The effects of sexual abuse on children. In *Lasting effects of sexual abuse*, (ed. G. E. Wyatt and E. J. Powell). Sage, Beverley Hills.

Conte, J. R., Wolf, S., and Smith, T. (1989). What sexual offenders tell us about prevention strategies. *Child Abuse and Neglect*, **13**, 293–301.

Corwin, D. L. (1988). Early diagnosis of child sexual abuse. In *Lasting effects of sexual abuse*, (ed. G. E. Wyatt and E. J. Powell). Sage, Beverley Hills.

Creighton, S. J. (1985). An epidemiological study of abused children and their families in the United Kingdom between 1977 and 1982. *Child Abuse and Neglect*, **9**, 441–8.

Cupoli, J. M. and Sewell, P. N. (1988). 1059 children with a chief complaint of sexual abuse. *Child Abuse and Neglect*, **12**, 151–62.

Dawson, J. (1989). When the truth hurts. *Community Care*. (30 March).

De Jong, A. R. (1989). Sexual interreaction among siblings and cousins: experimentation or exploitation? *Child Abuse and Neglect*, **13**, 271–9.

De Jong, A. R., Emmett, G. A., and Hervada, A. A. (1982). Epidemiologic factors in sexual abuse of boys. *American Journal of the Diseased Child*, **136**, 990–3.

De Mause, L. (1974). The evolution of childhood. In *The history of childhood*, (ed. L. De Mause). Psycho-History Press, New York.

de Young, M. (1988). The indignant page: techniques of neutralisation in the publications of paedophile organisations. *Child Abuse and Neglect*, **12**, 583–91.

Dimock, P. T. (1988). Adult males sexually abused as children. *Journal of Interpersonal Violence*, **3**, 203–21.

Dixon, K. N., Arnold, L. E., and Calestro, K. (1978). Father-son incest: underreported psychiatric problem? *American Journal of Psychiatry*, **135**, 835–8.

Dubé, R. and Hébert, M. (1988). Sexual abuse of children under 12 years of age: a review of 511 cases. *Child Abuse and Neglect*, **12**, 321–30.

Ellerstein, N. S. and Canavan, J. W. (1980). Sexual abuse of boys. *American Journal of the Diseased Child*, **134**, 255–7.

Emslie, G. J. and Rosenfeld, A. (1983). Incest reported by children and adolescents hospitalised for severe psychiatric disorder. *American Journal of Psychiatry*, **140**, 708–11.

Faller, K. C. (1989a). Why sexual abuse? An exploration of the intergenerational hypothesis. *Child Abuse and Neglect*, **13**, 543–8.

Faller, K. C. (1989b). Characteristics of a clinical sample of sexually abused children: how boy and girl victims differ. *Child Abuse and Neglect*, **13**, 281–91.

Fehrenbach, P. A., Smith, W., Monastersky, C., and Deisher, R. W. (1986). Adolescent sexual offenders: offender and offence characteristics. *American Journal of Orthopsychiatry*, **56**, 225–33.

Fehrenbach, P. A. and Monastersky, C. (1988). Characteristics of female adolescent sexual offenders. *American Journal of Orthopsychiatry*, **58**, 148–51.

Finkelhor, D. (1979). Sexually victimised children. Free Press, New York.
Finkelhor, D. (1984c). Four preconditions: a model. In *Child sexual abuse: new theory and research*, (ed. D. Finkelhor). Free Press, New York.
Finkelhor, D. (1984b). Boys as victims: review of the evidence. In: *Child sexual abuse: new theory and research*, (ed. D. Finkelhor). Free Press, New York.
Finkelhor, D. (1986). Abusers: special topics. In *A Source book on child sexual abuse*, (ed. D. Finkelhor). Sage, Beverley Hills.
Finkelhor, D. (1988). The trauma of child sexual abuse: two models. In *Lasting effects of child sexual abuse*, (ed. G. E. Wyatt and E. J. Powell). Sage, Beverley Hills.
Finkelhor, D. and Brown, A. (1986) Initial and longterm effects: a conceptual framework. In *A source book on child sexual abuse*, (ed. D. Finkelhor). Sage, Beverley Hills.
Freeman-Longo, R. (1986). The impact of sexual victimisation on males. *Child Abuse and Neglect*, **10**, 411–14.
Friedrich, W. N. (1988). Behaviour problems in sexually abused children: an adaptational perspective. In *Lasting effects of child sexual abuse*, (ed. G. E. Wyatt and E. J. Powell). Sage, Beverley Hills.
Friedrich, W. N., Beilke, R. L., and Urquiza, A. L. (1988). Behaviour problems in young sexually abused boys. *Journal of Interpersonal Violence*, **3**, 21–8.
Frisbie, L. V. (1969). *Another look at sex offenders in California*, Research monograph No. 12. California Department of Mental Hygiene, Sacramento.
Fritz, G. S., Stoll, K., and Wagner, N. N. (1981). A comparison of males and females who were sexually molested as children. *Journal of Sex and Marital Therapy*, **7**, 54–9.
Fromuth, M. E. and Burkhart, B. R. (1989). Longterm psychological correlates of childhood sexual abuse in two samples of college men. *Child Abuse and Neglect*, **13**, 533–42.
Furniss, T. H. (1990). Group therapy for boys. In *Working with sexually abused boys: an introduction for practitioners*, (ed. A. Hollows and H. Armstrong). TAGOSAC, London.
Furniss, T. H., Miller, L., and VanElberg, A. (1988). A goal-orientated group treatment for sexually abused adolescent girls. *British Journal of Psychiatry*, **152**, 97–106.
Gale, J., Thompson, R. J., Moran, T., and Sack, W. H. (1988). Sexual abuse in young children: its clinical presentation and characteristic patterns. *Child Abuse and Neglect*, **12**, 163–70.
Gebhard, P., Gagnon, J., Pomeroy, W., and Christenson, C. (1965). Sex offenders: an analysis of types. Harper and Row, New York.
Giaretto, H. (1981). A comprehensive child sexual abuse treatment programme. In *Sexually abused children and their families*, (ed. P. B. Mrazek and H. Kempe). Pergamon, Oxford.
Globe & Mail (1989). Orphanage complaints handled in 1982, (12 December). Toronto, 12 December, 1989.
Groth, N. and Burgess, A. (1979). Sexual trauma in the life histories of rapists and child molesters. *Victimology: An International Journal*, **4** 10–16.

Halpern, J. (1987). Family therapy in father-son incest: a case study. *Social Casework,* **68,** 88–93.

Haugaard, J. J. and Emery, R. E. (1989). Methodological issues in child sexual abuse research. *Child Abuse and Neglect,* **13,** 89–100.

Haynes-Seman, C. and Krugman, R. D. (1989). Sexualised attention: normal interraction or precursor to sexual abuse? *American Journal of Orthopsychiatry,* **59** 238–45.

Henderson, J. (1983). Is incest harmful? *Canadian Journal of Psychiatry,* **28,** 34–9.

Hobbs, C. L. and Wynne, J. M. (1987). Child sexual abuse: an increasing rate of diagnosis. *Lancet, ii,* 837–42.

Husain, A. and Chapel, J. L. (1983). History of incest in girls admitted to a psychiatric hospital. *American Journal of Psychiatry,* **140,** 591–3.

Ireton, H. R. and Thwing, E. J. (1974). Manual for the Minnesota Child Development Inventory. Behaviour Science System, Minneapolis, MN.

Johnson, R. I. and Shrier, D. (1987). Past sexual victimisation by females of male patients in an adolescent medicine clinic population. *American Journal of Psychiatry,* **144,** 650–2.

Johnson, T. C. (1988). Child perpetrators – children who molest other children: preliminary findings. *Child Abuse and Neglect,* **12,** 219–29.

Johnson, T. C. (1989). Female child perpetrators: children who molest other children. *Child Abuse and Neglect,* **13,** 571–85.

Jones, R. J., Gruber, K. J., and Timbers, G. D. (1981). Incidence and situation factors surrounding sexual assault against delinquent youths. *Child Abuse and Neglect,* **5,** 431–40.

Justice, B. and Justice, R. (1979). *The broken taboo.* Human Sciences Press, New York.

Kaufman, J. & Zigler, E. (1987). Do abused children become abusive parents? *American Journal of Orthopsychiatry,* **57,** 186–92.

Kavoussi, R. J., Kaplan, M., and Becker, J. V. (1988). Psychiatric diagnoses in adolescent sex offenders. *Journal of the American Academy of Child and Adolescent Psychiatry,* **27,** 241–3.

Keckley Market Research (1983). *Sexual abuse in Nashville: a report on incidence and longterm effects,* (March). Keckley Market Research, Nashville, TN.

Kercher, G. and McShane, M. (1984). The prevalence of child sexual abuse: victimisation in a adult sample of Texas residents. *Child Abuse and Neglect,* **8,** 495–502.

Kiser, L. J. et al. (1988). Post traumatic stress disorder in young children: a reaction to purported sexual abuse. *Journal of the American Academy of Child and Adolescent Psychiatry,* **27,** 645–9.

Kohan, M. J., Pothier, P., and Norbeck, J. S. (1987). Hospitalised children with history of sexual abuse: incidence and care issues. *American Journal of Orthopsychiatry,* **57,** 258–64.

Kolko, D. J., Moser, J. T., and Weldy, S. R. (1988). Behavioural/emotional indicators of sexual abuse in child psychiatric inpatients: a controlled comparison with physical abuse. *Child Abuse and Neglect,* **12,** 529–41.

Krug, R. S. (1989). Adult male report of childhood sexual abuse by mothers:

case descriptions, motivations and longterm consequences. *Child Abuse and Neglect,* **13**, 111–19.

Langsley, D. G., Schwartz, M. N., and Fairbairn, H. (1968). Father-son incest. *Comprehensive Psychiatry,* **9** 218–26.

Lewis, D. A., Shankok, S. S., and Pincus, J. H. (1979). Juvenile male sexual assaulters. *American Journal of Psychiatry,* **136** 1194–6.

Lewis, I. A. (1985) *Los Angeles Times,* poll No. 981, unpublished raw data.

Livingston, R. (1987). Sexually and physically abused children. *Journal of the American Academy of Child and Adolescent Psychiatry,* **26**, 413–15.

Longo, R. E. (1982). Sexual learning and experiences amongst adolescent sexual offenders. *International Journal of Offender Therapy and Comparative Criminology,* **26**, 235–41.

Longo, R. and Groth, A. N. (1983). Juvenile sexual offences in the histories of adult rapists and child molesters. *International Journal of Offender Therapy and Comparative Criminology,* **27**, 150–5.

McCarty, L. M. (1986). Mother-child incest: characteristics of the offender. *Child Welfare,* **65**, 447–58.

McCormack, A., Janus, M., and Burgess, A. W. (1986). Runaway youths and sexual victimisation: gender differences in an adolescent runaway population. *Child Abuse and Neglect,* **10**, 387–95.

McLeer, S. V., Deblinger, E., Atkins, M. S., Foa, E. B., and Ralphe, D. L. (1988). Post-traumatic stress disorder in sexually abused children. *Journal of the American Academy of Child and Adolescent Psychiatry,* **27**, 650–4.

Markowe, H. (1988). The frequency of child sexual abuse in the U.K. *Health Trends,* **20**, (1), 2–6.

Meiselman, K. (1978). *Incest: a psychological study of causes and effects with treatment recommendations.* Jossey-Bass, San Francisco.

Mian, N., Wehrspann, W., Klajner-Diamond, H., Le Baron, D., and Winder, C. (1986). *Child Abuse and Neglect,* **10**, 223–9.

Mohr, I. W., Turner, R. E., and Gerry, M. B. (1964). *Paedophilia and exhibitionism.* University of Toronto Press.

Mrazek, P. B. and Kempe, C. H. (1981). *Sexually abused children and their families.* Pergamon Press, Oxford.

Mrazek, T. J. and Mrazek, D. A. (1987). Resilience in child maltreatment victims: a conceptual exploration. *Child Abuse and Neglect,* **11**, 357–66.

Mrazek, P. B., Lynch, M., and Bentovim, A. (1981). Recognition of child sexual abuse In *Sexually abused children and their families,* (ed. P. B. Marazek and C. H. Kempe). Pergamon Press, Oxford.

Mullen, P. E., Romans-Clarkson, S., Walton, D. A., and Herbison, G. P. (1988). Impact of sexual and physical abuse on women's mental health. *Lancet,* (1) 841–5.

Murphy, J. E. (1985). Untitled news release (June) available from St Cloud State University, St Cloud, MN 56301.

Nasjleti, M. (1980). Suffering in silence: the male incest victim. *Child Welfare,* **59**, 269–75.

Peake, A. (1990). Under-reporting: the sexual abuse of boys. In *Working with sexually abused boys: an introduction for practitioners,* (ed. A. Hollows and H. Armstrong). TAGOSAC, London.

Peters, S. D., Wyatt, G. E., and Finkelhor, D. (1986). Prevalence. In *A sourcebook on child sexual abuse*, (ed. D. Finkelhor). Sage, Beverley Hils.

Pierce, L. H. (1987). Father-son incest: using the literature to guide practice. *Social Casework*, **68**, 67–74.

Pierce, R. and Pierce, L. H. (1985). The sexually abused child: the comparison of male and female victims. *Child Abuse and Neglect*, **9**, 191–9.

Pithers, W. D., Kashima, K. M., Cumming, G. F., and Beal, L. S. (1988). Relapse prevention: a method of enhancing maintenance of change in sex offenders. In *Treating child sex offenders and victims: a practical guide*, (ed. A. C. Salter). Sage, Beverley Hills.

Pomeroy, J. C., Behar, D., and Stewart, M. A. (1981). Abnormal sexual behaviour in prepubescent children. *British Journal of Psychiatry*, **138**, 119–25.

Raybin, J. B. (1969). Homosexual incest. *Journal of Nervous and Mental Disease*, **148**, 105–10.

Reinhart, M. A. (1987). Sexually abused boys. *Child Abuse and Neglect*, **11**, 229–35.

Righton, P. (1981). The adult. In *Perspectives on paedophilia*, (ed. B. Taylor). Batsford Academic.

Rimsza, M. E. and Niggemann, E. H. (1982). Medical evaluation of sexually abused children: a review of 311 cases. *Pediatrics*, **69**, 8–14.

Rogers, C. N. and Terry, T. (1984). Clinical interventions with boy victims of sexual abuse. In *Victims of sexual aggression: treatment of children, women and men*, (ed. I. R. Stuart and J. G. Greer). Van Nostrand Reinhold, New York.

Rosenfeld, A. A. (1979). The clinical management of incest and sexual abuse of children. *Journal of the American Medical Asosociation*, **242**, 1761–4.

Rosenfeld, A. A., Bailey, R., Siegel, B., and Bailey, G. (1986). Determining incestuous contact between parent and child: frequently of children touching parents' genitals in a non-clinical population. *Journal of the American Academy of Child Psychiatry*, **25**, 481–4.

Rosenfeld, A. A., Siegel, B., and Bailey, G. (1987). Familial bathing patterns: implications for cases of alleged molestation and for pediatric practice. *Pediatrics*, **79**, 224–9.

Russell, D. E. H. (1983). The incidence and prevalence of intrafamilial and extrafamilial sexual abuse of female children. *Child Abuse and Neglect*, **7**, 173–46.

Russell, D. and Finkelhor, D. (1984). Women as perpetrators: review of the evidence. In *Child sexual abuse: new theory and research*, (ed. D. Finkelhor). Free Press, New York.

Ryan, G. (1989). Victim to victimiser. *Journal of Interpersonal Violence*, **4**, 325–41.

Ryan, G., Lane, S., Davis, J., and Isaac, C. (1987). Juvenile sex offenders: development and correction. *Child Abuse and Neglect*, **11**, 385–95.

Sansonnet-Hayden, H., Haley, G., Marriage, C., and Fine, S. (1987). Sexual abuse and psychopathology in hospitalised adolescents. *Journal of the American Academy of Child and Adolescent Psychiatry*, **26**, 753–7.

Schecter, M. D. and Roberge, L. (1976). Sexual exploitation. In *Child abuse*

and neglect: the family in the community, (ed. R. Helfer and C. H. Kemp). Ballinger, Cambridge, MA.
Sebold, J. (1987). Indicators of child sexual abuse in males. *Social Casework*, **68**, 75–80.
Seghorn, T. K., Prentky, R. A., and Boucher, R. J. (1987). Childhood sexual abuse in the lives of sexually aggressive offenders. *Journal of the American Academy of Child and Adolescent Psychiatry*, **26**, 262–7.
Shane, P. G. (1989). Changing patterns among homeless and runaway youth. *American Journal of Orthopsychiatry*, **59**, 208–14.
Shoor, N., Speed, M. H., and Bartelt, C. (1966). Syndrome of the adolescent child molester. *American Journal of Psychiatry*, **122**, 783–9.
Singer, K. I. (1989). Group work with men who experienced incest in childhood. *American Journal of Orthopsychiatry*, **59**, 468–72.
Singer, M. I., Petchers, M. K., and Hussey, D. (1989). A relationship between sexual abuse and substance amongst psychiatrically hospitalised adolescents. *Child Abuse and Neglect*, **13**, 319–25.
Sirles, E. A., Smith, J. A., and Kusama, H. (1989). Psychiatric status of intrafamilal child sexual abuse. *Journal of the American Academy of Child and Adolescent Psychiatry*, **28** 225–9.
Smith, H. and Israel, E. (1987). Sibling incest: a study of the dynamics of 25 cases. *Child Abuse and Neglect*, **11**, 101–8.
Smith, M. and Grocke, M. (1990). Self concepts and cognition about sexuality in abused & non-abused children: an experimental study.
Spencer, N. J. and Dunklee, P. (1986). Sexual abuse of boys. *Pediatrics*, **78**, 133–8.
Sroufe, L. A. and Ward, M. J. (1980). Seductive behaviour of mothers of toddlers: occurrence, correlates and family origins. *Child Development*, **51**, 1222–9.
Steele, B. F. (1986). Notes on the lasting effects of early child abuse throughout the life cycle. *Child Abuse and Neglect*, **10**, 283–91.
Steele, B. F. and Alexander, H. (1981). Longterm effects of sexual abuse in childhood. In *Sexually abused children and their families*, (ed. P. B. Mrazek and C. H. Kempe). Pergamon, Oxford.
Stein, J. A., Golding, J. N., Siegel, J. M., Burnam, M. A., and Sorenson, S. B. (1988). Longterm psychological sequelae of child sexual abuse. The Los Angeles Epidemiological Catchment Area Study. In *Lasting effects of child sexual abuse*, (ed. G. E. Wyatt and E. J. Powell). Sage, Beverley Hills.
Stiffman, A. R. (1989). Physical and sexual abuse in runaway youths. *Child Abuse and Neglect*, **13**, 417–26.
Summit, R. C. (1983). The child sexual accommodation syndrome. *Child Abuse and Neglect*, **7**, 177–93.
Terr, L. C. (1987). Severe stress and sudden shock – the connection. Sam Hibbs Award Lecture, American Psychiatric Association Convention, Chicago, IL. (Reported in L. J. Kiser *et al.* (1988). Post-traumatic stress disorder in children: reaction to purported sexual abuse. *Journal of the American Academy of Child and Adolescent Psychiatry*, **27**, 645–9.)
Tong, L., Oates, K., and McDowell, M. (1987). Personality development following sexual abuse. *Child Abuse and Neglect*, **11**, 371–83.

Tufts' New England Medical Centre, Division of Child Psychiatry (1984). *Sexually exploited children: service and research project*, Final report for the Office of Juvenile and Justice and Delinquency Prevention. US Department of Justice, Washington DC.

Vander Mey, B. J. (1988). Sexual victimisation of male children: a review of previous research. *Child Abuse and Neglect*, **12**, 61–72.

Vizard, E. (1986). *Self esteem and personal safety*. Tavistock, London.

Wheeler, R. J. and Berliner, L. (1988). Treating the effects of sexual abuse on children. In *Lasting effects of child sexual abuse*, (ed. G. E. Wyatt and G. J. Powell). Sage, Beverley Hills.

Wild, N. J. and Wynne, J. N. (1986). Child sex rings. *British Medical Journal*, **293**, 183–5.

Wyatt, G. E. and Powell, G. J. (1988). Identifying the lasting effects of child sexual abuse: an overview. In *The lasting effects of child sexual abuse*, (ed. G. E. Wyatt and G. J. Powell). Sage, Beverley Hills.

Yates, A. (1982). Children eroticised by incest. *American Journal of Psychiatry*, **139**, 482–5.

Yates, A. and Terr, L. (1988a). Anatomical correct dolls: should they be used as the basis for expert testimony? *Journal of the American Academy of Child and Adolescent Psychiatry*, **27**, 254–7.

Yates, A. and Terr, L. (1988b). Issue continued: Anatomically correct dolls: should they be used as the basis for expert testimony? *Journal of the American Academy of Child and Adolescent Psychiatry*, **27**, 387–8.

4
Male rape in institutional settings
Michael B. King

It is well recognized that homosexual activity is more likely to occur in settings in which men are deprived of their usual heterosexual outlets. People are capable of a broad range of sexual response and many are aware of a homosexual component to their sexuality. The controversial work of Kinsey et al. (1948, 1953) first established in the public and scientific mind the definite possibility of a continuum of sexual response. In a study of 3392 women and 3849 men in North America, 50 per cent of male and 28 per cent of female subjects reported awareness of erotic responses to members of the same sex. This evidence supported a universal gradation between homosexuality and heterosexuality and, despite the obvious limitations of self-report, much subsequent work has confirmed the concept (McConaghy 1987).

In this chapter I am not concerned primarily with the variety of consensual sexual behaviour that may occur under the unusual conditions of institutional life, but more particularly with situations in which men are sexually assaulted. Prison assaults will be considered at greatest length, as this is the area about which most has been written and reported. However, sexual assaults in other institutional settings will also be considered.

Prisons

Although sexual assault is acknowledged to occur in prisons, surprisingly little research has been undertaken to delineate the nature and extent of the problem. This is due to the reluctance of many men to report that they have been sexually violated (Rideau and Sinclair 1982), as well as to the difficulty of undertaking this type of research in conservative institutional establishments. Prison authorities often dispute claims that there is a high level of sexual violence in institutions, but with little evidence one way or the other (Gunby 1981).

Nature and circumstances of the assaults

Sexual assault in prison is closely associated with humiliation and domination of the victim. Although the motive may not be entirely sexual (Rideau and Sinclair 1982), reports would indicate that the assault may form an initial mastery of a younger, weaker prisoner who, beaten and humiliated, later passively serves as an object of sexual gratification for the aggressor or aggressors. It is difficult to separate sexual arousal from a desire for domination and aggression, but there is little doubt that sexual gratification plays a greater role in coercive sexual activity in prisons than in the community, when assailants are rarely deprived of consensual sexual outlets. In a personal testimony of rape and violence in prisons, Tucker (1982) asserts that there is little to distinguish voluntary from coercive sexual activity among men in custody. It is a mistake to assume that so-called 'voluntary' or 'consensual' sexual relationships in this environment are completely free of aspects of power and control. Violence may not occur, simply because the victim is terrorized into submitting to the assailant's demands (Scacco 1982). Many relationships between men in prison may simply reflect the type of heterosexual relationship in which the active partner would engage in civilian life. Patterns of dominance that some men usually assert over female partners may be recast by domination of more submissive men in the sharply defined prison hierarchy.

Rideau and Sinclair (1982) characterized the pattern of sexual assaults that are meted out in some American prisons to newer prisoners, particularly those less able to defend themselves, many of whom later become the 'property' of the assailant. Men may be beaten and subjected to oral and anal penetration, as well as suffering humiliation in other ways, such as urination over their bodies or in their mouths. Victims may become sexual slaves who are made available to other prisoners in exchange for money or other barter. Assailants preserve their sense of masculinity by emasculating their victim, who subsequently accepts the psychological and physical pressure which dictates that, once emasculated, he must accept the passive role. In resigning himself to this manner of existence, the victim receives a degree of protection by his partner.

Herein lies a form of victimization that is peculiar to institutional life. Unlike the victim in the community, a raped prisoner must repay his rapist for the violence perpetrated on him by dedicating himself to serving his assailant's needs for perhaps years thereafter (Rideau and Sinclair 1982. Men who have been deprived of most avenues of self-expression and who have lost status by the act of imprisonment may resort to the use of sexual and physical power to reassert their uncertain male credentials (Gunby 1981). In this setting it is vital that the victim is feminized and that his attacker's sense of masculinity is preserved or even heightened.

There is some evidence that racial factors may be more important in prison assault than in assaults taking place in the community. In a small, controlled study of inmates identified and segregated by prison staff after having raped other prisoners, Moss et al. (1979) reported that there was a 'strong propensity' for black prisoners to fill the role of aggressor and whites that of the victim. Over one year, all 12 inmates isolated were black or of Mexican descent, while all but two victims were white. Similar findings had been reported several years earlier (Davis 1968; Carroll 1974; Scacco 1975) and the conclusion reached in these reports was that coerced sexual assault in all-male institutions may be a part of an ethnic power struggle taking place at a personalized level. A recent review of the subject of sexualized racial aggression has also supported this American finding of a disproportionate number of black aggressors and white victims, whether assaults take place in prisons or in juvenile corrective-training schools (Scacco 1982). The conclusion reached was that such black aggression reflected a deep-seated resentment harboured by lower-class blacks against middle-class whites.

Although others have also claimed that this racial bias underscores the aggressive, in contrast to sexual, nature of the attacks (Moss et al. 1979), it is difficult to reconcile why black or 'coloured' men are predominately reported to be the aggressors. The explanation may lie in the fact that black youths and young men are more 'street-wise' and better organized into groups within which their power is exercised than their white counterparts. It must be stressed, however, that the literature on this subject may be influenced by racial prejudice on the part of writers themselves (Buffum 1972).

Prevalence of sexual assault

Reluctance to report sexual assaults means that the extent of the problem is likely to be underestimated (Davis 1982). Many of the victims are distrustful of prison or legal authority and cannot take the step of making a complaint. There is also evidence that victims may be regarded by the authorities as somehow culpable in having attracted the assault (Scacco 1982), a familiar pattern in the rape of men or women in the community (Mezey and King 1989). Prevalence of sexual activity and assault in prisons depends to some extent on the level of control of prison violence by staff. In less well-regulated institutions in which prisoners have little recourse to protection, or in which there may be collusion between dominant prisoners and staff to maintain the peace, sexual violence tends to be greater (Davis 1982 Rideau and Sinclair 1982). Overcrowded conditions in which men have much idle time may also contribute (Gunby 1981).

Available figures from the study of prisoners themselves, in contrast to official records, suggest that between 19 and 45 per cent of men have had homosexual experiences while in prison in the United States (Buffum 1972) and at least 9 percent have been sexually assaulted (Gunby 1981). Other experts have claimed that homosexual activity is almost universal in some institutions and that sexual violence may be more prevalent as a result (Rideau and Sinclair 1982). Davis (1968) found that sexual assault in those awaiting trial in the Philadelphia prison system was almost 'epidemic'.

There is little information on the prevalence of sexual assault in British prisons. Levels of violence are, in the main, lower in British than American prisons and thus sexual assault may also be correspondingly less common. In the absence of well-conducted empirical research, however, this conclusion must remain based on speculation and anecdotal report.

Consequences of sexual assault

The consequences of sexual assault or coercive sexual relationships are severe, no less within prison than without. It has been claimed that violence erupting as a consequence of sexual coercion or homosexual activity may contribute significantly to acts of homicide within prisons (Gunby 1981). These may constitute killings in self-defence, or more commonly the results of a severe attack. Physical injury is not uncommon after gang assault (Elam and Ray 1986) and treatment may be the only hope of escape for the victim from the persecuting environment (Rideau and Sinclair 1982). Transmission of infectious diseases including hepatitis B virus and human immunodeficiency virus (HIV) are dangers in these assaults where there is anal or oral trauma. Until recently, sexually transmitted diseases have been the concerned mainly of those treating female victims (Jenny et al. 1990), but it is clear that men who have been raped are at risk (Hillman et al. 1990). The incarceration of drug users, together with their continued use of intravenous drugs while in prison, encurs a substantial risk of HIV transmission within that setting (Farrell and Strong 1991).

Psychological effects can only be guessed at in the face of little empirical data, either in Western Europe or the United States. A combination of physical and sexual assault with little recourse to justice, humiliation by both the attacker and by others in the prison, who quickly learn of the assault, and the requirement for continued submission to sexual demands constitutes a formidable pressure and it would seem inherently likely that severe psychological reactions result. Although suicide in prison may be a sequel of sexual assault (Wiggs 1989), there is little unequivocal evidence of an association (Salive et al. 1989). More is known of the psychological effect of male rape occurring in institutions other than prison and these will be discussed more fully below.

Although most victims survive in and adapt to this 'sexual jungle' (Rideau and Sinclair 1982), it is unclear how such experiences affect readjustment to a heterosexual lifestyle on release. Consensual homosexual activity within all male institutions appears to present little problem for later reassertion of a heterosexual lifestyle after release from custody. Thus, brief periods of homosexual behaviour would, in this sense, appear to be truly facultative. There may be a difference, however, in the case of coercive relationships where men are made to feel that their manhood has been taken from them and that they have been, in a certain sense, feminized. In a study in the 1970s, Sagarin (1976) examined ex-prisoners whose initiation into homosexuality had occurred during incarceration. He characterized three types of prisoners. First were so-called 'effeminates', men who were self-identified homosexuals before imprisonment and who took advantage of prison life to have sexual relationships. Second were 'involuntary recruits', who were raped, cajoled, threatened, or in some other way forced to submit to sexual relations with other male prisoners. Third, the 'aggressors' were those men who carried out the assaults. They were often older and stronger and made a heterosexual identification. These somewhat coarse stereotypes accord with other descriptions of prison sexual life. In Sagarin's view, consensual, non-threatening, or threatened male couples were rare in prison life because of the 'macho' requirements of that environment and the assumed inferiority of men 'choosing' affectionate homosexuality. Whether one inserts one's penis or becomes an insertee has important implications in prison life, in that the latter is viewed as submissive, inferior, and feminine. Once cast in that role, it appears to be almost impossible to lose the reputation (Sagarin 1976).

What becomes of involuntary recruits and aggressors after they leave prison? In this study of only nine men, Sagarin has claimed that aggressors quickly returned to heterosexual activities after release and viewed their prison conquests as submission of men who at some level desired the contact with them. This rationalization is not dissimilar to that used by men who are rapists of women in the community. The involuntary recruits, all of whom reported heterosexual lifestyles before incarceration, were subsequently living homosexual lives on release. Sagan concluded that this was evidence for the 'malleability' of sexual orientation.

Although addressing an interesting and important question, this work was seriously flawed in reporting on so few subjects and in recruiting the involuntary recruits from homosexual social circles. It is included here to indicate that while such work is possible, the quality of research conducted thus far provides very little understanding of the sexual experiences of men sexually victimized in prison. Work related to male victims in the community has indicated that, although men often experience some difficulty in sexual functioning, there is little lasting effect on their sexual orientation (see Chapter 1). The long-term impact on men who have been sexually victimized and humiliated over extensive periods in prison, however, without recourse to assistance either within prison or upon release, is quite unknown.

Sexual assault in other institutional settings

There is evidence that sexual assault of men may also occur in other institutional settings such as military establishments. Although these environments are less closed than that of prisons, the confines of institutional life appear to make sexual assault more likely and less easy to avoid. Although anecdotal reports of such assaults are not uncommon, there has been very little careful study of the extent of assault in such settings.

In a study based in a psychiatric out-patient clinic serving a population of Navy and Marine Corp men in the United States, Goyer and Eddleman (1984) reported that sexual assault appeared to be more common in this military setting than in other non-institutional settings. They described in some detail the effects of the assaults on 13 male victims, concluding that most men experienced mood disturbances, problems in relationships with peers, and sexual difficulties as sequelae to the assaults. Subsequent interpersonal difficulty was experienced, particularly as discomfort in being in close proximity to groups of men, and suspicion and uneasiness in sharing bathing and living quarters. Half those who had been molested wanted to be discharged from the military directly as a consequence of the assault.

Although not all these attacks occurred within the military establishment, several of them involved groups of assailants in assaults that took on something of the nature of initiation ceremonies. Similar themes emerged in the reports of these victims to those raped in prison, namely submission in the face of threatened or actual physical violence, humiliation, and fear of reporting the offence. One man believed that his sexual orientation had been fundamentally altered as a consequence of repeated assault. Goyer and Eddleman were struck by the profound and far-reaching effect such sexual assaults had on the victim's intrapsychic development, and on his social and psychological interactions with peers.

Conclusions

A greater understanding of the nature of sexual life in institutions and the circumstances that lead to assault is vital, but will only take place if authorities in charge of the institutions acknowledge the problem and allow research to proceed. From the evidence that we have to date, it is clear that sexual assault in prisons reflects patterns of violence, sexual aggression, and racism in the wider community. The hierarchy of male dominance, coupled with a restriction on all heterosexual expression, facilitates a pattern of assault and victimization of weaker or more isolated individuals. Coercive sexual activity is primarily an expression of anger and frustration in men who may be unable to achieve masculine identification and pride through avenues other than sex. More strict enforcement of discipline against offenders, although welcome, is unlikely to change the root cause of assaults, which lies in the frustrations of a class of men who seldom have work, successful families, or opportunities for emotional expression. Furthermore, the institutional structure may predispose to the problem. Without humanitarian changes within institutions the problem is unlikely to be reduced.

References

Buffum, P. C. (1972). *Homosexuality in prison*, US Department of Justice, Law Enforcement Assistance Administration, National Institute of Law Enforcement and Criminal Justice. Government Printing Office, Washington DC.

Carroll, L. (1974). *Hacks, blacks and cons: race relations in a maximum security prison*. D. C. Heath, Lexington, MA.

Davis, A. J. (1968). Sexual assaults in the Philadelphia prison system and sheriff's van. *Transaction*, **16**, cited in Davis A. J. (1982)

Davis, A. J. (1982). Sexual assaults in the Philadelphia prison system and sheriff's vans. In *A casebook of sexual aggression*, (ed. A. M. Scacco), pp. 107–20. AMS Press, New York.

Elam, A. L. and Ray, V. G. (1986). Sexually related tauma: a review. *Annals of Emergency Medicine*, **15**, 576–84.

Farrell, M. and Strong, J. (1991). Drugs, HIV, and prisons. *British Medical Journal*, **302**, 1477–8.

Goyer, P. F. and Eddleman, H. C. (1984) Same-sex rape of nonincarcerated men. *American Journal of Psychiatry*, **141**, 576–9.

Gunby, P. (1981). Sexual behaviour in an abnormal situation. *Medical News*, **245**, 215–20, 220.

Hillman, R. J. Tomlinson, D. McMillan, A., French, P. D., and Harris, J. R. W. (1990). Sexual assault of men: a series. *Genitourinary Medicine*, **66**, 247–50.

Jenny C. J. et al. (1990). Sexually transmitted diseases in victims of rape. *New England Journal of Medicine*, **322**, 713–16.

Kinsey, A. C., Pomeroy, W. B., and Martin, C. E. (1948). *Sexual behaviour in the human male*. W. B. Saunders, Philadelphia.

Kinsey, A. C., Pomeroy, W. B., Martin, C. E., and Gebhard, P. H. (1953). *Sexual behaviour in the human female*. W. B. Saunders, Philadelphia.

McConaghy, N. (1987). Heterosexuality/homosexuality: dichotomy or continuum? *Archives of Sexual Behaviour*, **16**, 411–24.

Mezey, G. and King, M. (1989). The effects of sexual assault on men: a survey of 22 victims. *Psychological Medicine*, **19**, 205–9.

Moss, C. S., Hosford, R. E., and Anderson, W. R. (1979). Sexual assault in a prison. *Psychological Reports*, **44**, 823–8.

Rideau, W. and Sinclair, B. (1982). Prison: the sexual jungle. In *A casebook of sexual aggression*, (ed. A. M. Scacco), AMS Press, New York.

Sagarin, E. (1976). Prison homosexuality and its effect on post-prison sexual behaviour. *Psychiatry* **39**, 245–57.

Salive, M. E., Smith, G. S., and Brewer, T. F. (1989). Prison rape and suicide – in reply. *Journal of the American Medical Association* **262**, 3403.

Scacco, A. M. (1975). *Rape In prison*. Thomsa, Springfield, Ill.

Scacco, A. M. (1982). The scapegoat is almost always white. In *A casebook of sexual aggression*, (ed. A. M. Scacco), pp. 91–103. AMS Press, New York.

Tucker, D. (1982) A punk's song: view from the inside. In *A casebook of sexual aggression*, (ed. A. M. Scacco), pp. 58–79. AMS Press, New York.

Wiggs, J. W. (1989). Prison rape and suicide. *Journal of the American Medical Association*, **262**, 3403.

5
Surviving sexual assault and sexual torture
Stuart Turner

Introduction

Sexual violence may occur in many settings – in established relationships or between strangers – but in its nature it is an attack that always challenges intimacy and trust. The specific psychological reactions, post-traumatic stress disorder and other sequelae, must be seen in the context of more pervasive and potentially more fundamental changes in which, for example, the nature of human relationships is laid bare for re-examination and re-evaluation. Although in this chapter, the basic reactions to psychological threat and trauma will be examined in more detail, the other personal sequelae should not be seen as less important.

Dealing with everyday traumas

Before turning to an examination of reactions to extreme violence, it may be helpful to consider how people deal with more common psychological challenges. Take, for example, the trivial case of a missed appointment. The first reaction derives from the mismatch that exists between internal expectations (cognitive assumptions about the external world) and the external reality. As time passes, so it becomes more and more unlikely that the person will arrive. This inevitably leads to a need to correct these internal expectations to provide a better fit with the external world. However, this simple mechanism is insufficient to deal with all the changes. In parallel with this cognitive processing of information, there is often a need to process the emotional reactions to this mismatch (Horowitz 1976).

Emotional processing is not restricted to unpleasant reactions. The person who believes that he or she has failed an examination will have some happy emotions to deal with when the results arrive and the prediction is disproved. Even in the example of a missed appointment, it is not too hard to conceive of situations in which the dominant reaction is relief rather than disappointment! It is possible to describe, in simple terms, some of the determinants of the nature of the emotional reaction, for example how many other tasks were facing the person and how much he or she had wanted the meeting.

After minor traumatic events, the emotional reaction is usually unpleasant. It may include anxiety, anger, disappointment, sadness, or misery. This condition of emotional distress is also dealt with by emotional processing. The person becomes aware of an emotional reaction and, through a process akin to habituation, the reaction slowly subsides. This process may be delayed in some people, for example by the demands of other daily activities or the overwhelming nature of the trauma. If this happens, then at some later stage, either through internal prompting or through an external cue reminding the person about the trauma, the emotional distress returns intrusively and can be processed.

Where the trauma is more extreme, it may be necessary to go through an avoidance and intrusive recall cycle many times before the emotional reaction is absorbed. After normal bereavement, the process of grieving (a concept akin to emotional processing) may take many months and the external cue of an anniversary may still cause a return of distressing memories. What happens if the trauma is even more extreme, outside the range of usual human experience?

Emotional processing and extreme trauma

The assumption presented here is that emotional processing is a normal adaptive mechanism by which major events (which lead to greater psychological arousal and subsequent upheaval) are assimilated. As the traumatic content increases, so more and more work may be required to deal with the emotional reaction. This is achieved through the phasic intrusion and avoidance cycle. Each intrusive recollection leads to the processing of only part of the emotional reaction before being replaced by an avoidance condition but eventually the matter is resolved. The emotional stimulation is broken down into manageable amounts before being processed over a period of time.

There are two exceptions to this simple rule. The person who manages to achieve effective avoidance may manage without the pain of repeated re-experiencing of the trauma. This is not the same as emotional processing. Coping by avoidance or distraction is not a complete solution and if this strategy breaks down later in life, perhaps at the time of a critical life change, there may be a full return of the intrusive experiences and associated distress.

Second, in extreme trauma, this normal process may be overwhelmed, resulting in a phasic reaction but one in which the emotional reaction is so extreme that it cannot be tolerated and the only effective solution is to enter an avoidance state. Emotional processing does not occur, or if it does it is insufficient to make much of an impact on the emotional disturbance. The consequence is that a cyclical intrusive-avoidance state persists for long periods.

Post-traumatic stress disorder

Post-traumatic stress disorder (PTSD) has recently entered the list of psychiatric diagnoses in the American 'Diagnostic and statistical manual'. In its current form (American Psychiatric Association 1987), PTSD is characterized by five elements (Table 1). The exposure to a major stressor, the intrusive recollections, the avoidance reactions, the state of increased arousal, and the persistence for at least one month may all be understood in terms of the failure of emotional processing.

Table 1 *Diagnostic criteria for post-traumatic stress disorder (DSM-IIIR*; abbreviated)*

A. The person has experienced an event that is outside the range of usual human experience and that would be markedly distressing to almost anyone.

B. The traumatic event is persistently re-experienced in at least *one* of the following ways:

(1) recurrent and intrusive distressing recollections of the event (in young children repetitive play in which themes or aspects of the trauma are expressed);
(2) recurrent distressing dreams of the event;
(3) sudden acting or feeling as if the traumatic event were recurring.
(4) intense psychological distress at exposure to events that symbolize or resemble an aspect of the traumatic event, including anniversaries of the trauma.

C. Persistent avoidance of stimuli associated with the trauma or numbing of general responsiveness (not present before the trauma), as indicated by at least *three* of the following:

(1) efforts to avoid thoughts or feelings associated with the trauma;
(2) efforts to avoid activities or situations that arouse recollections of the trauma;

(3) inability to recall an important aspect of the trauma;
(4) markedly diminished interest in significant activities (in young children, loss of recently acquired developmental skills);
(5) feeling of detachment or estrangement from others;
(6) restricted range of affect (e.g. unable to have loving feelings);
(7) sense of a foreshortened future.

D. Persistent symptoms of increased arousal (not present before the trauma) as indicated by at least *two* of the following:

(1) difficulty falling or staying asleep;
(2) irritability or outbursts of anger;
(3) difficulty concentrating;
(4) hypervigilance;
(5) exaggerated startle response;
(6) physiologic reactivity upon exposure to events that symbolize or resemble an aspect of the traumatic event.

E. Duration of the disturbance (symptoms in B, C and D) of at least one month.

---

* DSM-IIIR, Diagnostic and Statistical Manual (3rd edition, revised) (American Psychiatric Association 1987).

Typical post-traumatic reactions (Ramsay 1990) are characterized by intrusive thoughts, memories, and even vivid flashbacks in which the person has the experience of reliving the traumatic episode. These processes often affect sleep, with both increased sleep latency and a very restless and disturbed sleep pattern. Sometimes, sleep is broken by shouts and signs of high autonomic arousal. There are also avoidance reactions, including both internal mechanisms such as emotional numbing and avoidance of external cues that remind the person of the trauma. This type of condition has been described after a wide range of personal tragedies, including war trauma, torture, and major disaster (Wilson and Raphael 1991). While representing an extreme end of a range of responses from adaptive to maladaptive, to some degree it typifies the reaction to all personally traumatic events.

PTSD and personal vulnerability

There has long been a tendency to report psychological disturbance after trauma in terms of a personal weakness or vulnerability. For example, after the First World War, shell shock, if not perceived as an organic brain reaction to severe explosions, was often regarded as a manifestation of cowardice and personal failure. Serious psychological disturbance and handicap are often ignored in assessments of the legitimacy of compensation claims. The psychoanalytic perspective, in which adult personal difficulties are understood in terms of childhood conflicts reactivated, was misguidedly used to deny the validity of the genuine and appalling psychological trauma many had experienced.

In the literature on disasters, the debate continues between those who find that the amount of trauma is critical (e.g. Shore et al. 1986) and those who find that personality type is of primary importance (e.g. McFarlane 1988) in determining post-traumatic reactions.

Rachman (1980) reviews the evidence that different components of the trauma may lead to failure of emotional processing. Events that are sudden, intense, dangerous, uncontrollable, unpredictable, and irregular are seen as more likely to overwhelm normal processing than events that are signalled, safe, mild, controllable, predictable, and progressive. Similarly, events that constitute 'prepared' stimuli (Rachman 1980), that it to say, stimuli to which people naturally seem to be more likely to develop phobic reactions, such as snakes and spiders, are more damaging.

Rachman also lists several personality variables that have been shown to play some part in mediating the degree of failure in emotional processing. These include neurotic and introverted personality traits.

How many these apparently conflicting formulations be understood? The obvious solution is to see these factors not as competing but as interacting. In this way, a moderate trauma to a sensitive person may lead to high level of distress with PTSD. On the other hand, a more hardy person may require a larger trauma to show a similar reaction. The phrase 'everyone has his breaking point' is derived from the experiences of soldiers in the First World War (Horowitz 1976). Anyone exposed to a high enough level of trauma will react with incomplete emotional processing. In other words, at high levels of trauma, personal differences become irrelevant; at moderate levels of trauma, they may be more important mediating variables.

Rachman also recognizes that associated factors such as high arousal, illness, fatigue, sleeplessness, intense concentration, heat, noise, and stressor overload may be significant in determining the final reaction.

PTSD and sexual violence

The particularly intrusive nature of sexual violence places it within Rachman's list of stimulus factors causing difficulty in processing. In a society in which men are expected to defend themselves and passivity is equated with femininity (Mezey and King 1987), a failure to repel a sexual assault becomes a mark of masculine inadequacy and therefore may be an injury too threatening to reveal. Power and aggression, rather than simple sexual gratification, are the dominant motivations for sexual assault on men or women. The act is intended to produce submission and humiliation, and the challenge to the sexuality and potency of the victim is therefore paramount.

Torture may include overt sexual assault or a more covert attack on a victim's sexual integrity. It serves to illustrate some of the processes that follow other forms of male sexual assault. Torture has been practised in a widespread way in one in every three countries in the 1980s, affecting many millions of people. It may be defined as the use of physical or psychological pain to achieve the purposes of the torturer over the will of the victim. Its purpose is clear and always involves a power relationship, with subjugation of the victim a primary aim and sexual assault one of the ways of achieving this.

The United Nations (1984) definition makes it clear that torture is always carried out by or at the instigation of a public official. It is in its nature, therefore, the act of a state against an individual, an attack by the dominant on the dependent within a society. Many of the psychological processes exemplified in torture are also seen in other forms of assault that do not involve political acts but in which there is a similar abuse of a dependent relationship.

The stimulus characteristics, as listed by Rachman, that militate against successful processing are systematically used by the torturers to achieve their effects. In all cases, the purpose of torture is for one group or community to achieve domination over another. It is hardly surprising that sexual assault, with its background of aggression, humiliation, and domination, should constitute an important part of this process. The torture may range from enforced nakedness with sexual threats or humiliations through electrical shocks and physical violence to genitals, finally to anal rape with batons, actual sexual rape by the torturers, or even partial or complete castration.

Agger (1988) reports a unique body of information compiled by political prisoners in El Salvador. Forty torture methods are listed in a book of testimonies collected in prison by interviewing 434 fellow prisoners between February and August 1986, and recording their torture experiences. Six of the torture methods included a form of sexual assault. These were blows to the testicles, electrical torture, nakedness, rape, threats of rape, and other methods including physical damage to genitals. Seventy-six per cent of the men reported that they had been subjected to at least one of these methods.

The sexual trauma is often too painful for the person even to think about and, although survivors of torture will often talk in a therapeutic setting about many aspects of their experiences, sexual assault is a very private matter which may only be mentioned after many sessions, if at all. If it is so difficult to describe, the effort involved in processing the experience must be very great indeed. In a recent investigation of 100 survivors of torture, although most forms of trauma were significantly associated with marked intrusive features, sexual torture was unique in that it showed a significant relationship with avoidance symptoms alone (R. Ramsay personal communication).

There are several possible explanations for this reluctance to talk about sexual assault. It may be that the circumstances of the interview are inhibiting. In a large community survey (the Los Angeles Epidemiologic Catchment Area Project), in which both male and female interviewers were employed, the rates of reported sexual assault tended to be higher, both for men and for women, when interviewed by female interviewers (Sorenson et al, 1987). It is equally possible that sexual assault is a subject about which even trained interviewers show some avoidance, in a manner similar to the 'conspiracy of silence' between the terminally ill and their health care professionals (Hinton 1967). A further major factor in reluctance to disclose sexual assault is the distress experienced by the victim when trying to remember the assault.

Shame is an important clue to understanding the severity of this reaction and the reluctance of survivors to disclose sexual trauma. One of the processes common in torture and present in other assaults, is the victim's belief that, somehow, he is responsible for his own pain. The sexual trauma may be coupled with deep personal guilt, which leads him to anticipate criticism from others. With this in mind, one of the important therapeutic processes must be to re-establish that the trauma was externally motivated, and not the personal responsibility of the victim.

By causing profound shame and guilt, sexual torture leaves the person passive in both the private and socio-political areas of life and therefore serves its violent and repressive purposes (Agger 1988). By attacking private fears of homosexuality, the perpetrator of other forms of sexual assault both humiliates and isolates his victim, sometimes leading to long-term and undisclosed suffering. Friends suspecting problems are also likely to choose avoidance rather than confrontation of the pain of abuse.

One of the stated purposes of the book of testimonies was to serve as an accusation against the Salvadorean authorities. However, it seems probable that the collective statement also had profound effects on each individual's understanding of the torture process. By reframing the individual memories within a political context it helped survivors to place their experiences in a more accurate and acceptable light.

Secondary victimization

In the same way that the perpetrator seeks to leave his victim with the belief that the assault was invited, so afterwards others may give the same impression. Survivors of sexual assault risk vilification by family, friends, and, ultimately, the police and judicial system. A process of justice that rests on adversarial conflict may result in enormous distress for victims of assault and reopen psychological wounds.

Perceived negative attitudes to the man who has experienced sexual assault may reactivate the trauma experience. Bearing in mind the two underlying psychological themes of castration anxiety and homosexual anxiety, described by Agger (1988), it is not hard to see how these may be affected by post-assault experiences, for example the common tendency to assume that victims of male sexual assault must be overt or secret homosexuals. With the increasing prevalence of human immunodeficiency virus (HIV) infection and AIDS in many parts of the world, the concern about sexually transmitted disease becomes another important element, and occasionally an excuse for prejudice, in defining society's attitude to victims of sexual assault.

These ambivalent attitudes are critical barriers to the individual who needs to talk about what has happened. Such negative responses are being challenged in some areas, notably female rape and in reactions to natural disasters, but there is much still to do. It is possible that some of the anti-male rhetoric associated with female rape crisis centres, in which men are cast only in the role of aggressor, is serving to add to the isolation of the male victim of sexual assault (Mezey and King 1989).

Completing the work of emotional processing

Successful emotional processing may allow a return to undisrupted behaviour as emotional disturbance declines. For the person with PTSD, this may only happen after a therapeutic intervention. Theoretically, the essential element in the facilitation of emotional processing is the re-experiencing of the emotional reaction in such a way that the individual does not respond by avoidance and numbing. To achieve this, the therapy must be calm and unhurried and must allow the person plenty of time to go over what has happened. The therapeutic 50 minutes may be insufficient and some workers have used flexible scheduling with early sessions lasting for two or three hours. There is a wide body of opinion from many fields, although little evaluative research, that favours the rehearsal of the trauma story as an essential element of treatment. However, agitated rehearsal or brief recollections without the opportunity for habituation may be harmful. It may sensitize the survivor even more strongly to the traumatic event.

Sexual disturbance

After torture and sexual assault, sexual problems are common. In a study on 17 Greek men, tortured between 1967 and 1975, reduced libido and erectile dysfunction was reported in 29 per cent (Lunde *et al.* 1980, 1981). Levels of luteinizing hormone, follicle-stimulating hormone, prolactin, and testosterone were no different from controls but that sexual dysfunction was more common in those who had suffered genital truama. This demonstrated that sexual dysfunction after torture was related to the psychological meaning of the trauma rather than hormonal changes.

In the main, reports on torture survivors have not included information about sexual dysfunction or sexual torture (Rasmussen 1990). In one large study with a systematic approach to record-keeping, the history of sexual trauma was omitted in more than half the cases (Rasmussen 1990). These findings appear to indicate that avoidance behaviour affects therapists as well as survivors. It is important to consider the possibility of sexual assault in patients presenting with sexual dysfunction and a thorough assessment of this possibility is always required.

For those survivors with PTSD, it is common to find that events which resemble the precipitating trauma reactivate the original distress. Whatever the trigger, they may lead to high levels of tension and subsequent sexual dysfunction. Sexual activity with a partner may reactivate memories of the trauma by association. Explanation and appropriate reassurance may be very powerful therapeutic approaches.

Depression

After any form of extreme trauma, there may be associated losses. For some this may be loss of health, occupation, relationship, or the ability to trust. After any of these traumatic losses, there may be a despressive reaction. In investigation of survivors of torture and war trauma, depression frequently coexists with PTSD (Turner and Gorst-Unsworth 1990, 1991). It may range in severity from pessimism and social withdrawal to a profound depression with weight loss, sleep disturbance, and even, rarely, a psychotic reaction.

Existential aspects

Severe trauma of any sort will produce some psychological change. For the majority this will be negative but in a few people the crisis may paradoxically lead to increased self-confidence and self-regard. The effects on personal attitudes and values may be least easy to describe in generalities. These almost certainly depend on personal experience and development. After torture, for example, it has been reported in a series of 37 survivors that 32 per cent had difficulties in establishing new relationships and 38 per cent had an inability to trust other individuals (Randall *et al.* 1985). These general statements are likely to hide a complex variety of individual responses in which personal meanings and value systems, religious and political notions, are challenged by the disturbing reality of the trauma experience.

Effects on others

Finally, it is important to consider the effects on partners and spouses of survivors of sexual assault. Not only do they have to manage their own traumatic reaction, they may also find themselves in the position of having to offer support to the nominal victim. After natural disasters, the wide-ranging impact has now been accepted. Taylor (1989) lists five types of victim in addition to those people who were actually present at the time. He also includes their families and friends, and the emergency and counselling services, as well as three other miscellaneous groups. Where the counselling of survivors of sexual assault, takes place within lay counselling groups specializing in this work, the possibility that counsellors may become victims should be an important consideration in assessing the need for external support and supervision.

Conclusion

In this chapter, the characteristic psychological reactions to male sexual assault in the context of torture have been outlined. Victims of violence are too often seen as in some way unusual or defective if they show their psychological distress. If the effects of victimization are to be reduced, the first step must always be to place these psychological conditions within the range of understandable reactions to major external stressors. This often means that a significant conflict must be addressed openly. On the one hand, a survivor of an assault may have a right to compensation or asylum, and this will depend on a satisfactory medical report indicating the presence of an illness condition. On the other hand, the survivor may be reluctant to accept that they have a psychological disorder with all the associated stigma. By explaining that psychological reactions are more like injuries than diseases, and by being open about the way these reactions develop, it may be possible to reconcile these apparently opposing perspectives.

It is also important to adopt a whole-person approach. As this chapter has illustrated, there may be a range of interacting reactions affecting psychological, sexual, and other physical aspects of the individual. The assault will have taken place within a socio-political environment. To ignore any one of these aspects leads to incomplete or inadequate formulation and approach to therapy.

Finally, it is likely that, as in other forms of trauma exposure, individuals who have survived the experience may be able to offer the most useful assistance. Principles of self-help, informed by professional expertise, may be important elements in any therapeutic response.

References

Agger, I. (1988) *Psychological aspects of torture. Psychological aspects of torture with special emphasis on sexual torture: sequels and treatment perspectives*. Paper presented to the WHO Advisory Group on the Health Situation of Refugees and Victims of Organised Violence (Gothenburg).

American Psychiatric Association (1987). *Diagnostic and Statistical Manual of Mental Disorders, (3rd ed, revised)*. APA, Washington DC.

Hinton, J. (1967). *Dying*. Penguin, London.

Horowitz, M. J. (1976). *Stress response syndromes*. Jason Aronson, New York.

Lunde, I., Rasmussen, O. V., Lindholm, J., and Wagner, G. (1980). Gonadal and sexual functions in tortured Greek men. *Danish Medical Bulletin*, 27, 243–5.

Lunde, I., Rasmussen, O. V., Wagner, G., and Lindholm, J. (1981). Sexual and pituitary-testicular function in torture victims. *Archives of Sexual Behavior*, 10, 25–32.

McFarlane, A. C. (1988). The aetiology of post-traumatic stress disorders following a natural disaster. *British Journal of Psychiatry*, 152, 116–21.

Mezey, G. and King, M. (1987). Male victims of sexual assault. *Medicine Science and the Law*, 27, 122–4.

Mezey, G. and King, M. (1989) The effects of sexual assault on men: a survey of 22 victims. *Psychological Medicine*, 19, 205–9.

Rachman, S. (1980). Emotional processing. *Behaviour Research and Therapy*, **18**, 51–60.
Randall, G. R. *et al.* (1985). Physical and psychiatric effects of torture (US study). In *The breaking of bodies and minds*, (ed. E. Stover and E. O. Nightingale). Freeman, New York.
Ramsay, R. (1990). Post-traumatic stress disorder; a new clinical entity? *Journal of Psychosomatic Research*, **34**, 355–65.
Ramsay, R., Gorst-Unsworth, C., and Turner, S. W. (1991). Psychological reactions to torture: a large retrospective series. *British Journal of Psychiatry* (submitted for publication).
Rasmussen, O. (1990). Medical aspects of torture. *Danish Medical Bulletin*, **37** (suppl. 1), 1–88.
Shore, J. H., Tatum, E. L., and Vollmer, W. M. (1986). Psychiatric reactions to disaster: the Mount St Helens experience. *American Journal of Psychiatry*, **143**, 590–5.
Sorenson, S., Stein, J. A., Siegel, J. M., Golding, J. M., and Burnam, M. A. (1987). The prevalence of adult sexual assault: the Los Angeles Epidemiologic Catchment Area Project. *American Journal of Epidemiology*, **126**, 1154–64.
Taylor, J. A. W. (1989). *Disasters and disaster stress*. AMS Press, New York.
Turner, S. W. and Gorst-Unsworth, C. (1990). Psychological sequelae of torture: a descriptive model. *British Journal of Psychiatry*, **157**, 475–80.
Turner, S. W. and Gorst-Unsworth, C. (1991). Psychological sequelae of torture. In *The international handbook of traumatic stress syndromes*, (ed. J. Wilson and B. Raphael) Plenum Press, New York (in press).
United Nations (1984). *Convention against torture and other cruel, inhuman or degrading treatment or punishment*. Office of Public Information, United Nations, New York.
Wilson, J. and Raphael, B. (1991). *The international handbook of traumatic stress syndromes*. Plenum Press, New York (in press).

6
Male co-survivors: the shared trauma of rape
Daniel C. Silverman

Rape precipitates a shared life-crisis in its primary and secondary survivors. A striking paradox is that those people closest to the female survivor, the secondary survivors, (in particular, the boy-friends, husbands, brothers, and fathers) may at first be unable to provide her with the understanding and support she so desperately needs because of their own complex cognitive and affective responses to rape. Although the discussion here considers the male partners of female rape survivors, it is likely that much of what follows applies as well to the woman's female family members or to female partners of men who have been raped.

The purpose of this chapter is to describe the acute cognitive and affective responses of male co-survivors of rape, to identify patterns of psychological and behavioural response and their underlying psychodynamics in co-survivors, to suggest techniques for clinical interventions with secondary survivors in the acute aftermath of rape, and to consider the special challenges for male counsellors involved in rape-trauma intervention work with female survivors and their male partners.

Rape: a shared life-crisis

One boy-friend of a survivor who had been raped some days earlier did not learn of it from his girl-friend until a troubled attempt at love-making. He immediately began to experience symptoms. At first he described himself as feeling 'complete shock and disbelief that what I was hearing could possibly be true!'. After a few hours, he felt intense anxiety and helplessness and didn't know how to comfort the girl-friend. 'I was completely overwhelmed. I couldn't think of what to do or say. I felt panicked. Then suddenly I remembered that a woman we both knew had been raped a year before and received counselling in a rape crisis programme.' Although it was after one o'clock in the morning he urged the girl-friend to get dressed and rushed her across town by car so that she could speak to the friend who would 'know all the right things to say'. As soon as he had arrived at the friend's apartment and left the women talking, he began to ruminate about the details of the rape experience, including the fact that the rapist had forced the girl-friend to perform oral sex. At this point he experienced a feeling of intense physical disgust that paralleled the girl-friend's initial reaction. 'I threw up and felt ashamed at how strong my own responses were. I felt like I had really lost control of myself.'

Extensive clinical experience has shown that the emotional crisis of rape traumatization is a mutual one shared by the survivor and those closest to her (Silverman 1978; Holmstrom and Burgess 1979; McCartney 1980; Hertz and Lerer 1981; White and Rollins 1981; Orzek 1983). Male partners frequently describe an inescapable feeling that they 'have been raped too'. In the immediate post-traumatic period, partners of the survivor commonly describe experiencing shock, disbelief, guilt and shame, overwhelming anxiety and helplessness, intrusive thoughts, rage, and physical revulsion. These cognitive, affective, and physiological reactions mirror the most common immediate responses of the rape survivor herself, previously described as central aspects of the initial phase of the 'rape trauma syndrome' (Sutherland and Scherl 1970; Burgess and Holmstrom 1974). Such responses clearly embody the range of reaction patterns from psychological numbing and denial to the hyper-arousal seen in the immediate post-traumatic reactions of the primary survivor. These reactions are identical to those of the primary rape survivor and represent post-traumatic stress responses occurring in the secondary survivor. More enduring aspects of post-traumatic stress disorder (American Psychiatric Association 1987) can also be seen in secondary survivors and include emotional lability or numbing, depression, loss of interest in normal activities, withdrawal from intimate relationships, and sexual dysfunction.

The shared crisis of rape can stress all aspects of relationships. Rape creates a situation where both the strengths as well as the vulnerable aspects of relationships are heightened. The capacity of a relationship to withstand stress before the rape will be the best predictor of a couple's ability to cope with the impact of the rape experience. Changes in the interpersonal dynamics of couples and families occur directly as a result of the intrapsychic disequilibrium experienced by the primary and secondary survivors (Silverman 1978; Holmstrom and Burgess 1979; McCartney 1980; Hertz and Lerer 1981; White and Rollins 1981; Orzek 1983). The crisis of rape is certain to accentuate pre-existing attitudes, expectations, and problems concerning the giving and receiving of care, sex-role responsibilities, and sexual contact (Holmstrom and Burgess 1979).

The sudden and shocking nature of rape makes it impossible for those closest to the survivor to prepare for the crisis it precipitates

(Bard and Ellison 1974). By contrast, 'developmental' crises are more gradual in onset and allow partners, spouses, and family members more time to adapt to changing life-circumstances. The acutely disorganizing impact of rape upon relationships in the areas of independence and interdependence, sexuality, intimacy, trust, and the expression of anger may be immediately apparent. The age and developmental stage of the survivor, her partner, and family members determine which psychological functions will be affected by the rape. Developmental areas affected may include those of the normal emergence of mature sexuality, attainment of a separate adult identity, and achievement of a balance between autonomy and mutual dependence (Notman and Nadelson 1976; Hertz and Lerer 1981; White and Rollins 1981).

What follows deals largely with the impact of the rape experience upon co-survivors in the immediate post-traumatic period. Clinicians offering long-term care to survivors of rape and their partners report an extensive range of relationship difficulties, often marked by disturbances in sexual functioning that persist long after the acute post-traumatic period (Becker et al. 1982; Miller et al. 1982; Orzek 1983) More consideration of the particular problems in communication and sexual intimacy that rape precipitates in couples will be found below in the section (Care and over-protection) that deals with patterns of response in co-survivors.

Myths, misconceptions, and misalliances: patterns of cognitive response in co-survivors

Although one might expect people closest to the survivor to provide unconditional emotional support, this does not take into account the complexity of the emotional response to rape trauma. The coexistence of violent and sexual elements makes for an incongruity peculiar to rape traumatization. Paradoxically, those closest to the rape survivor, particularly the men in her life, may find it difficult to offer unequivocal support because of ambivalent attitudes, misconceptions, and beliefs held concerning the emotionally laden matter of rape.

In addition to shared psychological and physiological responses, male co-survivors may manifest certain patterns of cognitive reactions to rape in the immediate post-traumatic period. Co-survivors are subject to the same misapprehensions, myths, and prejudices that inevitably seem to surround the crime of rape (Amir 1971; Morrison 1980). Perhaps the most common bias is that of conceptualizing rape solely as a sexual experience rather than a violent and life-threatening event with sexual elements. Other familiar misconceptions concerning rape include ideas such as 'nice girls don't get raped', 'women who get raped are asking for it', 'women who want to get raped show it by dressing in a provocative manner', and 'there is no such thing as being raped by a husband, boy-friend, date, or acquaintance; rape can only happen between strangers'. Attitudes such as these may be a reflection of male-dominated cultures but they also betray a latent anger toward the woman for allowing herself to become victimized. Such resentment is usually not experienced consciously at first by the co-survivor and is rarely expressed directly. Unconscious anger with the survivor is the most serious impediment to the mobilization of unconditional support from co-survivors. If left unaddressed it can be destructive to relationships.

The co-survivor may resent the survivor for 'allowing this awful thing to happen' and in so doing, precipitating a life-crisis for the co-survivor marked by feelings of guilt, inadequacy, suffering, shame, and self-doubt. 'Blaming the victim' is a common human response to emotional crisis and distances the co-survivor from feelings of responsibility and self-criticism. The important matter of helping co-survivors resolve these difficult feelings will be addressed in detail below in the section dealing with clinical aspects of rape counselling. Co-survivors' anger is often expressed in subtle and oblique ways. For example, it is not uncommon for boy-friends, husbands, and fathers to display doubts about the survivor's version of events, particularly in cases of rape by a 'date' or acquaintance. They may ask the woman to repeat her story over and over again while searching for small inconsistencies or things the survivor could have done differently.

The husband of a woman who was raped by a local tradesman in a vacant lot 20 yards from the town's police station was incredulous that his wife had not been able to summon help. His wife tried to explain that the combination of her own terror, the attacker's threat to hurt her if she cried out, and the fact that he held his hand over her mouth during most of the rape made it impossible to draw attention to herself. In spite of this explanation, the husband repeatedly asked the wife how long the rape had gone on and expressed disbelief 'that in that amount of time it was impossible to get help if she had really wanted to'.

Indirect anger may be expressed in the form of criticisms which imply that the woman should have been more careful, even in situations where it is clear that carelessness played no role in the rape.

One man's girl-friend, who lived in a safe neighbourhood, was raped by an intruder who entered her fourth-floor apartment through a narrow bathroom window as she slept. The boy-friend criticized the woman for not having the foresight to install iron window grates.

A husband whose wife was assaulted as she walked home along her usual route felt that she should have noticed that the neighbourhood had been changing for some time and arranged a regular ride home from work with friends.

Some male co-survivors may regard the woman as their personal property (Brownmiller 1975). Paternalistic attitudes concerning the expression of female sensuality, veneration of virginity, and a sense of exclusive ownership of the woman's sexuality may be related to construing the rape experience as a sexual event having taken place outside the man's exclusive relationship with 'his woman'.

Male partners may feel personally wronged by the assault and display a proprietary sense of outrage. Another source of anger directed at the survivor may be the man's perception that the rape represents a 'breach of sexual fidelity'. This anger may be more of a defence mechanism against his sense of vulnerability, fear of rejection by the woman, and loss of masculine self-esteem than a reflection of deeply held, chauvinistic convictions.

A man discussing the basis of his emotional upset about his girl-friend's rape stated: 'It's as if, for want of a better term, my property had been transgressed against.' Another male co-survivor feared losing his partner's interest because, 'she had such an exciting sexual experience with a powerful man'. One husband commented after learning of his wife's rape, 'I felt so betrayed. It was as if she had chosen to have an affair with another man.'

Closely related feelings that may interfere with supportive and compassionate responses on the part of male partners involve anger that the woman has allowed herself to become 'damaged merchandise' or irreversibly 'tainted' by the rape. Hidden feelings, including the fear that the woman will be stigmatized, psychologically damaged, or made 'frigid', may cause the male co-survivor to withdraw emotionally from the woman when unconditional support is most critical. His unconscious dilemma is a shared sense of devaluation and feared loss of self-worth.

One boy-friend of a rape survivor expressed concern that rape would leave indelible emotional scars and described a fantasy of seeing 'a scarlet "R" burned into her forehead. Although I know it's not true, it's as if she's changed somehow, as if she was spoiled by it.' Another man lamented the fact that all the women with whom he became emotionally involved seemed to suffer from psychological hang-ups. After asking, 'Why me?' he said, 'This one was fine until the rape, now I'm sure she'll be like all the rest!'

It should be clear from the foregoing that a response of unconditional support and empathy may be difficult if the co-survivor is burdened by his own post-traumatic stress reaction as well as the common misapprehensions that surround rape. For these reasons, involvement of the co-survivors in clinical interventions is vital to avoid compounding the trauma of the primary survivor. Secondary survivors may need assistance in expressing their troubling ideas about rape in an accepting environment, before a process of education and counselling can begin. Both individual and conjoint counselling interventions may be indicated in many cases. Aspects of clinical interventions with co-survivors, such as the counsellor's possible counter-transference reactions to co-survivors' unempathic responses, are discussed in more detail in the section below concerning the male counsellor.

'Care and over-protection': patterns of emotional and behavioural response in co-survivors

Male co-survivors may react in a variety of predictable ways to counter feelings of powerlessness, guilt for having failed to protect the woman, or unconscious resentment toward the survivor. In their desire to assist the survivor and to contain their own feelings of helplessness, they may attempt to mobilize the involvement and support of the survivor's family, women friends, professional counsellors, clergymen, teachers, and colleagues. In the immediate post-rape period, the woman may experience this as intrusive and overwhelming, particularly if she feels that her privacy and autonomy are being threatened.

Tolerating the feelings associated with rape trauma can be very difficult for the male co-survivor and may lead to a range of counterproductive coping behaviours. Taking control of the situation, infantilization, and over-protection frequently emerge (Silverman 1978). Attempts to move the survivor as far away from the original setting of the rape as possible, accompanying her to and from home and in all of her daily routines, and keeping her under constant surveillance and supervision are common strategies. Distraction is another familiar tactic used by male co-survivors. Parties and get-togethers with friends, shopping sprees, and hastily arranged vacations are used in an urgent attempt to keep the survivor and her partner from dwelling on their thoughts and feelings about the rape.

Some male co-survivors encourage the woman to keep the rape a secret in order to 'protect' other family members from the anticipated trauma they would experience in learning the truth. Common examples of this are, 'Your mother's not strong enough to handle it'; 'It would kill your dad if he found out'; or 'They're just kids, they're too young to know or understand anything about it' in the case of survivors with children. The motivations for secrecy may vary but include parental discomfort with one's own or a child's sexuality, fear of being blamed for failing as a protector, intense shame, or long-standing family patterns of avoiding emotionally charged topics. The purpose of trying to hide the rape or distract oneself from it is to attempt to undo the effects of the rape, to function as if it never happened. Efforts at secrecy, distraction, and undoing grow out of the co-survivor's conviction that open-ended discussion of the trauma keeps alive destructive, disorganizing memories for the survivor and himself (Silverman 1978). While both primary and secondary survivors reach a point in the process of recovery from rape where they don't want to be reminded of what has occurred, a premature ending to the expression of feelings and distress may deny the survivor and those sharing in her crisis the opportunity to mourn the losses inherent in rape traumatization. This may also have the effect of depriving all survivors of a much-needed feeling of mutual support and can serve to confirm fears that what has happened is too awful to discuss.

Powerful fantasies of revenge against the rapist are another response seen on the part of many partners, brothers, and fathers of survivors. Male partners burdened by recurrent thoughts of retaliation may purchase weapons, make threats of violence, and search out the rapist. In contrast, overt expressions of anger by the female survivor are less common (Mezey and Taylor 1988). Such actions may serve to protect the man from directly experiencing the feelings of powerlessness, vulnerability, and impotent rage that he unconsciously shares with the primary survivor. In some cases, such fantasy and behaviour may be the means by which the male co-survivor experiences or communicates the woman's anger, which for reasons of cultural conditioning or fear of retaliation she is unable to express for herself.

One father of a rape survivor, overwhelmed with rage and uncontrollable fantasies of violent revenge, bought a semi-automatic assault rifle and began patrolling the rundown neighbourhood where his daughter had been raped in the hope of finding her assailant. He was arrested by the police for possession of an unlicensed weapon while driving alone in his car during the midnight hours. After weeks of anxiety concerning her father's safety, the daughter was finally able to prevail upon him to give up his nightly surveillance missions.

In extreme cases, the woman may find herself in the burdensome position of calming, placating, and restraining the men who would be her avengers. The man's over-zealousness can have the undesirable effect of adding to the survivor's burden and may place her in the stressful position of having to provide soothing, containment, and reassurance rather than receive it.

In the weeks and months after rape, problems related to communication and sexual functioning frequently emerge for survivors and their male partners (Miller et al. 1982; Cohen 1988). These disruptions in communication and intimacy hold the greatest threat to the stability of the relationship if left unresolved. Common patterns of disrupted communication may begin with the woman's desire to put the rape out of her mind. This leads to a tendency to avoid the expression of feelings or discussion of symptoms related to the rape. The man is left in the position of trying to determine what the woman wants and needs from him (Cohen 1988). As indicated earlier, the male co-survivor is burdened by his own feelings and post-traumatic symptoms, now compounded by his confusion as how best to help his partner. As the woman becomes increasingly dependent upon her partner for support, the man feels that he must not add to her distress by expressing his own needs. Over time this impossible situation leads to anger and resentment on the man's part, which, not surprisingly, increases his sense of guilt and confusion. As his anger increases, the male partner is less able to display patience with the survivor's mood changes, her need to be taken care of, ambivalence about physical affection, avoidance of sexual contact, and occasional displacement of anger from the rapist to her partner. Ultimately, the man's escalating resentment can lead to withdrawal of support, culminating in a sense of mutual alienation that further worsens communication.

The disordered patterns of communication contribute to the sexual problems commonly seen (Holmstrom and Burgess 1979; Becker et al. 1982; Miller et al. 1982). Some men may feel hesitant to ask for a return to the sexual relations that existed before the rape, fearing that this will re-traumatize the woman. After several days or weeks, the woman may be experiencing less acute anxiety and starting to regain her interest in sexual relations but avoids initiating them because of what she perceives as lack of interest on the man's part. Difficulty in communicating needs and desires openly, compounded by feelings of resentment, lead to a misunderstanding of partners' wishes and intentions.

Perhaps the most common and problematic aspect of sexual functioning after rape is the question of when to resume sexual relations (Holmstrom and Burgess 1979). As discussed in the section below on educating co-survivors, rape is primarily an act of violence. However, as sex is the 'weapon' with which the woman is attacked, survivors often have powerful fears about resuming sexual activity after the rape. Both the woman and her male partner may fear that sexual contact will bring back painful or frightening memories of the rape. Sexual dysfunctions, including fear of sex, inhibition of desire or arousal, are not uncommon (Becker et al. 1982). It appears that the survivor's physiological sexual responsiveness is less affected as a consequence of the rape than is the tendency to experience sexual stimuli as anxiety-provoking or to find one's sexual feelings lessened or inhibited.

Many male partners wish to resume sexual relations with the survivor as soon after the rape as is possible (Holmstrom and Burgess 1979). Motivations may include the man's wish to prove to himself and his partner that the rape has not changed anything, his need to reassert his 'ownership' of the woman, to protect against fears of losing the woman's sexual interest, or to convince himself that he can overcome his aversion to sexual relations with a raped, and therefore 'spoiled' or 'unfaithful', woman. Here again, the capacity for open communication or the degree to which it has been adversely affected by the rape will determine the degree of sexual dysfunction. The woman's sexual responsiveness after rape may be affected by whether she experienced the rape event as predominantly a sexual act or a physical assault, and the extent to which the rape was reminiscent of the couple's customary sexual practices. The couple's ability to decide together what are the limits of non-threatening and pleasurable sexual activity will determine how difficult the resumption of relations will be. It is in the man's best interest to allow the woman control over the sexual part of the relationship as to the degree of physical intimacy and its timing. Clearly, their ability to discuss the sexual part of their relationship in the broader context of the survivor's feelings about the rape, her need for support, affection, and patience from her partner, and the current state of her interest and readiness for sexual activity will determine when a mutually satisfying physical closeness can be achieved.

Ventilate then educate: counselling considerations

Counselling interventions must be designed to confront both the central fact and essential paradox of rape traumatization. Rape precipitates a mutual life-crisis in the primary survivor and those persons closest to her, and, initially, co-survivors may not be able to offer unconditional help because of their own reactions and responses to the shared trauma. In order to create a safe environment for the recovery of primary and secondary survivors of rape trauma, counselling interventions must provide the following:

1. Opportunity for survivors and partners to acknowledge the shared life-crisis precipitated by rape and to express their feelings in response to it in a setting that provides confidentiality, non-judgemental acceptance, and nonintrusive support.
2. Help for co-survivors in understanding the inherently terrifying, violent, and traumatizing nature of rape victimization for the pri-

mary survivor, which precipitates profound feelings of vulnerability, devaluation, and sense of loss of control over her life.

3. Education about the nature of rape-related, post-traumatic stress and its psycho-social and physiological sequelae in both primary and secondary survivors. These include periods of apparent outward adjustment, punctuated by the return of the rape experience in the form of flashbacks, nightmares and intrusive thoughts, physical symptoms, anxiety, depression, phobias, and feelings of emotional detachment.

4. Provision of counselling, educational and therapeutic services to primary and co-survivors of rape for those somatic, psychological, and interpersonal reactions that disrupt their ability to cope effectively and impede recovery from rape-related trauma.

Clinical experience with survivors of rape has made it clear that crisis counselling, educational, and psychotherapeutic work with the partners of rape survivors is often critical to a successful outcome for both primary and secondary survivors. An essential component of such interventions is helping the survivor and members of her support network to define the trauma of rape victimization as a mutual life-crisis. The conscious acknowledgement that rape precipitates a crisis for the woman as well as those people closest to her serves to reframe the dilemma in the broadest terms of shared concern and responsibility. It offers permission for co-survivors to recognize and acknowledge their own pain, stress, and symptoms with less guilt and confusion, and invites all parties to participate in the healing process. Recovery from rape trauma should not be conceptualized as solely the problem or responsibility of the primary survivor but rather the shared work of her wider social network.

As indicated earlier, even those people closest to the survivor may be subject to the pervasive myths, misunderstandings, and prejudices surrounding the crime of rape that exist in the community at large. While consideration of their underlying dynamics may make such responses more understandable, a critical dilemma is that, if not contained, these attitudes and their associated impact and behaviours make revictimization of the woman a real possibility. Counsellors must be vigilant to the presence of these common misconceptions and the feelings that accompany them. It is crucial that they be recognized and, with the assistance of the counsellor, openly discussed. Direct expression of these thoughts and affects will offer far greater control over counter-productive or ambivalent messages that can arise between survivor and partner.

It may be useful to offer male partners individual counselling in which negative attitudes about women who are raped can be identified and discussed. The counsellor should use his or her position as an authoritative but non-judgemental teacher with expertise in rape to disabuse partners and family members of their misconceptions. By offering each the privacy of separate meetings, difficult issues can be discussed and clarified in a confidential and safe environment.

Counsellors may have to cope with what they perceive as the unempathic, narcissistic, or chauvinistic responses of male co-survivors. Counsellors need to understand the origin of such apparently unenlightened attitudes. Implied criticism of the man for his apparently self-centred concerns may grow out of a failure to appreciate that he is also in a crisis and feels under assault by way of identification with his partner. It is necessary first to allow him to ventilate his deepest concerns and misapprehensions, free from fear of criticism or rejection by the counsellor. The counsellor must try to focus on and empathize with the man's pain and sense of loss. In addition, the counsellor can help the man understand that while rape is traumatic for the woman and those closest to her, survivors need not be permanently disabled by the experience. While there may always be painful memories attached to the rape, recovery is possible, with a return to a satisfying life.

A vital part of the counselling process with the family and partners of the rape survivor is education. The counsellor needs to teach the co-survivors about the nature of the rape experience as well as its predictable emotional and physical sequelae. Co-survivors must be helped to understand that, for the survivor, the experience has been one of exposure to violence, terror, dehumanization, and humiliation, rather than an erotic encounter. The common assumption that the rapist is motivated by sexual desire and that rape simply represents an attempt to gratify sexual needs has to be confronted. Clinical study has revealed that stereotypical portrayals of the rapist as merely sexually frustrated or as a sex-fiend with insatiable perverse drives are inaccurate (Amir 1971; Groth *et al.* 1977). Instead, rape is the behavioural expression of largely non-sexual motives. Rape represents a multiply determined act with sexual elements, which combines the expression of hostility towards women with a disturbed need to exert a sadistic form of power and control over them (Groth *et al.* 1977).

Understanding the kinds of feelings the woman is experiencing is an important beginning in fostering responses from co-survivors that support the woman's attempts to manage her own recovery. Clearly a woman who senses that her male partner fails to understand the most basic aspects of what she has been through will find it hard to experience him as genuinely empathic.

Survivors and their partners should be helped to prepare for the foreseeable psychological and somatic sequelae of rape known as the 'rape trauma syndrome' (Burgess and Holstrom 1974), a type of post-traumatic stress disorder (American Psychiatric Association 1987). Counsellors should alert survivors to the possible return of the rape experience in the form of recurrent and intrusive memories, dreams, nightmares, and flashbacks of the event. Survivors and partners can be familiarized with potential symptoms, such as emotional numbing, avoidance of stimuli that remind one of rape, sexual dysfunction, insomnia, irritability, emergence of phobias, hypervigilance, anxiety attacks, depression, crying and intense feelings of fear, anger, self-blame, and loss of self-esteem. Prior awareness of potential patterns of reaction may help survivors and their partners respond to the symptoms with less alarm, guilt, and helplessness. The need for stress reduction through the use of relaxation exercises, self-hypnosis, somatic treatments including the judicious use of anxiolytic medications, mutual support groups for survivors of rape, and longer-term psychotherapy for issues of self-blame and loss of self-esteem should be assessed during initial rape-crisis counselling interventions with primary and secondary survivors.

In the acute post-traumatic period, the counsellor should teach the co-survivors that they are most helpful when they allow the woman to use her own best coping capacities. Previously described patterns of infantilization and over-protection on the part of the survivor's partners, while well intentioned, run the risk of communicating to the woman that she is seen as a vulnerable child in need of protection and could reinforce her fears that she is indeed helpless. This may prevent her from using her most adaptive coping strategies at less cost to her self-esteem and sense of being an autonomous adult. The counsellor must intervene in a way that supports the co-survivor's understandable wish to be helpful, while at the same time helping him to see the potential disservice that can be done by fostering dependence in the rape survivor. Partners should be encouraged to allow the woman to express her feelings and needs openly without intervening precipitously to undo the pain and 'make everything all right'. Co-survivors need to learn that rape, like other life-crises, need not make it impossible for the woman to function adaptively in most areas of her life and that a return to a more normal existence is possible in time.

Perhaps the single most important message the counsellor can bring to the partners of the survivors is that of containment of difficult feelings. Here the task is one of explaining and modelling how to provide a safe and supportive 'holding' environment (Winnicott 1965) in which the woman and her co-survivors can release their most disturbing thoughts and feelings without fear of criticism or condemnation emerging from a shared sense of helplessness. The counsellor will need to display empathy, address sensitive and painful material, and refrain from being directive. The message to be conveyed is that the reality of rape cannot be undone. Then the task becomes one of grieving this fact while at the same time helping co-survivors realize and value that in being emotionally available, willing to hear the woman's concerns, and tolerating the feelings they share, they are offering their loved one much that is immediately helpful and that can ultimately be healing.

Special considerations: survivors of rape trauma and the male counsellor

Staffing demands for crisis intervention work in many institutions make it inevitable that male counsellors, mental health workers, psychologists, psychiatrists, and physicians will frequently be the first or only person available to the female rape survivor. It was believed in the beginning of the rape-crisis intervention movement, during the early 1970s, that women survivors would prefer to receive help only from women health-care providers (Bassuk et al. 1975; McCombie and Arons 1980). Clinical experience has shown that many female survivors are comfortable with and willing to speak to male rape-crisis counsellors and health-care providers (Silverman 1977). While all possible efforts should be made to honour the request of those survivors who feel they would prefer to speak to a female care-giver, women can find it relieving and reassuring to begin a process of reaffirming trust and confidence in the care-taking and compassion of men (Silverman 1977; Silverman et al. 1988).

Male counsellors may experience certain initial difficulties in providing compassionate and empathic care to rape survivors (Silverman 1977). Not surprisingly, these problems may parallel those discussed for the male co-survivors. Upon beginning the work of crisis intervention and counselling follow-up with rape survivors, the male counsellor may wonder if his help is wanted and may fear rejection by the woman survivor. He may assume that because he is a man he will be seen as a symbol of the male aggression and violence to which the survivor has been exposed. A powerful wish to 'make up' for the suffering that has been inflicted and simultaneously convince the woman that not all men need be feared or loathed may inform his actions. He may feel under pressure to turn the interaction into an 'emotionally corrective' experience by proving that some men, in the person of the counsellor, are truly sensitive, empathic, and trustworthy. Such unspoken agendas may explain why junior male psychiatric staff being trained as counsellors in a rape-crisis intervention programme described themselves as altering their usual clinical styles and techniques in the emergency room when dealing with survivors of recent rapes (Silverman 1977). Some counsellors modified the way in which they used space, physical contact, tone, and volume of voice. They also considered certain subjects inappropriate to discuss with the woman during interviews. Those counsellors who were very aware of and concerned about the sexual trauma inherent in the rape experience tended to keep a 'proper' distance from the woman, felt that any physical contact was inappropriate, and thought the sexual details of the rape and the woman's sexual history should be elicited only by the female medical nurse. Those male counsellors most concerned with the survivor's fearfulness, loss of self-esteem, and emotional upheaval tended to sit closer to the woman, speak to her in soft and reassuring tones, and felt it was appropriate to hold her hand or place their arm on her shoulder to offer comfort and support. If preconceptions about rape trauma are carried to an extreme, they may lead either to unnecessary distancing or potentially intrusive or seductive clinical interactions. No one strategy is appropriate for every case and changes in approach based upon unaddressed anxieties in the counsellor or biases in working with rape survivors may produce less authentic and sensitive clinical interventions.

Although male counsellors may be capable of great sensitivity and caring, initially they may experience more difficulty than their female counterparts in identifying with the woman rape survivor. Feelings of helplessness, loss of control, and vulnerability are rape survivors most universal responses. Male counsellors may find it difficult to fully empathize with such feelings if they unconsciously experience themselves as inept or helpless in trying to support the survivor through her overwhelming sense of crisis. Some counsellors may mobilize a defence of identification with the aggressor (the rapist) or the primary male partner (boy-friend, husband, or father). This can serve to protect against a sense of powerlessness to help the woman. Such defensive identifications may be manifested in doubting the credibility of the woman's story, feeling concerned about the accused man's rights, or sharing the male partner's anger with the woman for somehow 'allowing the rape to happen' and, in the process, damaging herself and her man.

While the male counsellor's identification with the survivor's boy-friend, husband, father, and brother may also serve as a basis for empathy with the co-survivor's experience, it can be problematical when unexamined identifications extend to commonly held male attitudes about women. Counsellors may find themselves experiencing indignation that the 'woman-as-an-extension-of-her-man' has been raped and by implication her man and all men have been wronged. Counsellors' unconscious anger toward certain survivors for apparently 'encouraging' rape, by means of inappropriate dress or behaviour for example, can lead to an accusatory style of reviewing the woman's history for clues that indicate her responsibility. This 'blaming the victim' may serve to absolve the men in her life of any responsibility or shared sense of blame.

A trainee counsellor was assigned the case of a young woman who was suddenly and brutally attacked from behind and raped as she walked alone to the bus stop by a man she had met earlier in the evening at a party. In the first session the woman described herself as 'having had what I thought was just a pleasant conversation' with this man. In supervision, the counsellor reported a strong impression that this woman 'brought the rape upon herself by exercizing poor judgement and probably having been flirtatious with the man without having realized it'.

The male counsellor who is unaware of his own prejudices, misconceptions, and identification with the male co-survivors may also be prone to some of their counter-productive strategies for helping the victim. Paternalistic over-protectiveness, patronizing attempts to rescue the woman from her painful feelings and distress through facile reassurance, or by being overly directive may impede the woman's recovery.

Male counsellors need to be sensitive to various aspects of the female survivor's psychological state in the immediate post-traumatic period. Because of the intense feelings of guilt, shame, devaluation, and vulnerability that survivors experience, beginning new relationships, particularly with a man, may be difficult. The female survivor may be cautious about entering into an alliance with the counsellor, no matter how sincere his intentions. The counsellor should not regard this as a personal rebuff or interpret it as a rejection of help in general. For these reasons, it is best that the counsellor offer help and availability in a gentle, non-insistent manner. No immediate or firm commitment to long-term counselling should be expected or sought from the survivor. An offer of telephone follow-up may be experienced as less intrusive and allowing the survivor more control over the frequency and intensity of future contacts. In this regard, another not uncommon feature of survivors' initial presentations may be the tendency to displace anger concerning the rapist or her current state of suffering. The woman may be afraid to express her rage with the rapist more directly because of fears of retaliation or a feeling of not having a right to be angry because she holds herself responsible for the rape. The anger may be redirected toward friends, family, police, and, not infrequently, helping professionals. Such a response can be confusing and discouraging to any well-meaning care-provider. Being able to recognize the woman's difficulty in expressing her rage towards her assailant or situation can allow the counsellor to respond to the anger in a non-defensive and non-retaliatory manner.

Research on the help-seeking patterns of women surviving different types of rape indicates that survivors who display a great deal of self-blame and guilt over what has happened may take longer to seek out medical and psychiatric care than survivors experiencing less self-blame (Silverman et al. 1988). Some survivors may feel that, because of their shame, guilt, and sense of responsibility for the rape, they do not deserve help. Still others wish to put the rape out of their minds as quickly as possible. The male counsellor must be careful to respect the survivor's need to control the amount of care she can tolerate and should avoid being too aggressive in trying to convince the woman to accept help.

One final factor that can affect the character of male counsellors' clinical work with female survivors may be a desire to please female colleagues or supervisors in the rape counselling setting. Attempts to prove oneself a politically correct and sensitive man, whose attitudes about rape and women's issues in general are liberal and liberated, can lead to overly solicitous behaviour with survivors, peers, and supervisors. Such zeal may only serve to protect the man and his female colleagues from a deeper consideration of the potentially troubling attitudes, feelings, and concerns that he brings to his rape counselling work.

References

American Psychiatric Association (1987). Post-traumatic stress disorder. In *Diagnostic and statistical manual of mental disorders*, (3rd edn, revised), pp. 247–51. APA, Washington DC.

Amir, M. (1971). *Patterns of forcible rape*. University of Chicago Press.

Bard, M. and Ellison, K. (1974). Crisis intervention and investigation of forcible rape. *Police Chief*, **41**, 70–5.

Bassuk, E., Savitz, R., McCombie, S., and Pell, S. (1975). Organizing a rape crisis intervention program in a general hospital. *Journal of the American Medical Women's Association*, **30**, 486–90.

Becker, J. V., Skinner, L. J., Abel, G. G., and Treacy, E. C. (1982). Incidence and types of sexual dysfunction in rape and incest victims. *Journal of Sex and Marital Therapy*, **8**, 65–74.

Brownmiller, S. (1975). *Against our will*. Simon and Schuster, New York.

Burgess, A. W. and Holmstrom, L. L. (1974). Rape trauma syndrome. *American Journal of Psychiatry*, **131**, 981–6.

Cohen, L. J. (1988). Providing treatment and support for partners of sexual assault survivors, *Psychotherapy*, **25**, 94–8.
Groth, A. N., Burgess, A. W., and Holmstrom, L. L. (1977). Rape: power, anger and sexuality. *American Journal of Psychiatry*, **134**, 1239–43.
Hertz, D. G. and Lerer, B. (1981). The rape family: family reactions to the rape victim. *International Journal of Family Psychiatry*, **2**, 301–15.
Holmstrom, L. L. and Burgess, A. W. (1979). Rape: the husband's and boyfriend's initial reactions. *The Family Co-ordinator*, **28**, 321–30.
McCartney, C. F. (1980). Counselling the husband and wife after the woman has been raped. *Medical Aspects of Human Sexuality*, **14**, 121–2.
McCombie, S. L. and Arons, J. H. (1980). Counselling rape victims. In *The rape crisis intervention handbook*, (ed. S. L. McCombie), pp 145–71. Plenum Press, New York.
Mezey, G. and Taylor, P. (1988). Psychological reactions of women who have been raped: a descriptive and comparative study. *British Journal of Psychiatry*, **152**, 330–9.
Miller, W. R., Williams, A. M., and Bernstein, M. H. (1982). The effects of rape on marital and sexual adjustment. *American Journal of Family Therapy*, **10**, 51–8.
Morrison, C. H. (1980). A cultural perspective on rape. In *The rape crisis intervention handbook*, (ed. S. L. McCombie), pp. 3–16. Plenum Press, New York.
Notman, M. and Nadelson, C. C. (1976). The rape victim: psychodynamic considerations. *American Journal of Psychiatry*, **133**, 408–13.
Orzek, A. M. (1983). Sexual assault: the female victim, her male partner, and their relationship. *The Personnel and Guidance Journal*, **62**, 143–6.
Silverman, D. C. (1977). First do no more harm: female rape victims and the male counsellor. *American Journal of Orthopsychiatry*, **47**, 91–7.
Silverman, D. C. (1978). Sharing the crisis or rape: counselling the mates and families of victims. *American Journal of Orthopsychiatry*, **48**, 166–73.
Silverman, D. C., Kalick, S. M., Bowie, S. I., and Edbril, S. D. (1988). Blitz rape and confidence rape: a typology applied to 1000 consecutive cases. *American Journal of Psychiatry*, **145**, 1438–41.
Sutherland, S. and Scherl, D. J. (1970). Patterns of response among victims of rape. *American Journal of Orthopsychiatry*, **40**, 503–11.
White, P. N. and Rollins, J. C. (1981). Rape: a family crisis. *Family Relations*, **30**, 103–9.
Winnicott, D. W. (1965). The theory of the parent-infant relationship. In *The maturation process and the facilitating environment*, pp. 43–6. International Universities Press, New York.

7
Cultural and historical aspects of male sexual assault

Ivor H. Jones

The few existing historical accounts of sexual assault on males are insufficient to form a reliable picture of attitudes and behaviours in earlier epochs but give some insight into contemporary attitudes of key cultures – Greco-Roman, medieval European, post-seventeenth century Europe can be found, with fleeting references to exotic cultures. Hints of behaviour in prehistory could come, by analogy, from contemporary hunter-gatherer groups, although search of Australian aboriginal literature has not produced much useful material. By adopting a socio-biological rather than an historical paradigm, and by examining not attitudes but behaviour seen in a ethnological context, hypotheses relevant to male rape can be examined – specifically, the proposition that, amongst primates, male mounting of the male is more a function of dominant – subordinate relationships than of their sexual relationships – and this theme appears to be congruent with many of the historical data. The evidence suggests that sexual assault of one male by another seems to be used for non-sexual, social purposes – usually control of one individual by another.

Definitions

In historical accounts, problems of definition abound; they arise in defining sexual assault as opposed to sexual co-operation and this ground can readily be shifted by legislative fiat. Male sexual assault is generally regarded as a subset of homosexual behaviour, with implications and thereby problems of definition extending beyond the genital act to sexual preference. Such a view is not in accord with contemporary accounts (Sykes 1958; Scacco 1981) but the assumption pervades the historical literature.

In Roman culture, for substantial periods, a homosexual act was not a matter of great concern but questions about who was the more passive and especially the social status of those involved in passive sexual acts were important. These cultural attitudes thereby coloured Roman accounts (Breiner 1990). According to Boswell (1980), in Greek literature it was not necessary to specify the erotic content of intense relationships between males, but it would have been surprising if eroticism were absent. There are also special problems: sexual behaviour was, until very modern times, frequently not described in anatomical or interactional terms but was alluded to loosely, often by literary or biblical allusions. Indeed, 'homosexual' is itself a recent term dating from the late nineteenth century. Kraft Ebbing used it, as did Havelock-Ellis, albeit without approval (Boswell 1980). Before this, a bewildering series of terms had been used to denote those who engaged in male/male sexual behaviour, including sodomite, invirion, homogenic, and even uranian. Sodomy was perhaps the most consistent, but commonly described the 'inappropriate vessel' rather than the gender of the participants.

A distinction has to be made between acts of violence with and without a sexual component. For example, Torquemada and the Inquisitors burned men alive as an *auto de fe* without an explicit sexual component. The Aztecs attempted to remove the heart from a living victim in honour of the god of war, Huitzilopochtli, again without an overt sexual component. But the infidels, for equally high reasons at the time of the Crusades, are reported to have tortured their victims by impaling them with a sharpened stake into the anus until, sometimes after a day or so, it appeared in the neck. In these situations, whether the act was sexual or not depends as much on the motivation and responses of the perpetrator as on the act itself, and generally the acts were justified on religious grounds rather than acknowledging any element of self-gratification.

Various initiation rites also fall into this category and generally have multiple functions. As well as the sexual component, other aspects of initiation, in particular of power and group conformity, may satisfy complex psychological needs for the participants. A contemporary initiation rite among Cornell (American university) freshmen is described in which the initiate is required to provide a lubricated 6–inch nail and, as part of the ceremony, to bend over in what amounts to a sexual presentation. Instead of the expected outcome, he is given a bottle of beer to drink and the celebrations commence (Tiger 1969).

The common practice of genital greasing among Australian apprentices may similarly be regarded as an initiation rite, but with strong sexual overtones. The act is performed by the use of force, penetration does not usually occur – but such behaviours are very close to male/male rape. Eibl Eibesfeldt (1971) cites other examples of initiation rites that are tantamount to male rape; for example, by Polish and

Hungarian shepherd boys who rape male strangers intruding into their territory.

Rome

Boswell (1980) provides the most detailed account of homosexual behaviour throughout almost 200 years of the Roman period. He quotes Edward Gibbon as saying that, of the first 15 Roman emperors, Claudius 'was the only one whose taste in love was entirely correct'. By repute, Roman sexual relations were frequently of a type that must now be considered as assault but this would be to misunderstand the prevailing ethos. Accounts of homosexual rape of Roman citizens were favourite themes in contemporary folklore. Disapprobrium was expressed, not against the homosexual activity, which, at least early in the Empire was widely tolerated, nor even against the use of force in certain social contexts (for example, citizen-slave), but against acts of passivity on the part of citizens. Boswell notes: 'It is extremely difficult to convey to modern audiences the absolute indifference of most Latin authors to the question of gender.' Nor was force the critical issue it is for us: passivity by a Roman citizen was the central issue. The concept of *stuprum*, meaning defilement, was commonplace, and this was concerned with passivity shown by a citizen (not a non-citizen) outside a socially approved context. Status was central to the context. Slaves, foreigners, or prostitutes could be used for sexual purposes without constraint. Seduction, rape, and prostitution of a free-born, including free-born minors, did constitute *stuprum*. Specifically excluded were acts committed against citizens involving *force majeur*, for example, sexual assault inflicted by brigands or enemies. The brigands and enemies were, of course, severely punished if caught, but social displeasure did not extend to the victims.

Rape of minors of either sex who were citizens was severely punished, seduction being viewed as a lesser crime. Homosexual and heterosexual acts of this nature were regarded equally. In this respect, Boswell notes that contemporary accounts, particularly of Greek origin, specifically criticize the Romans for limiting their eroticism to older youths.

Although expressions of distaste for homosexuality can be found in Roman literature, negative attitudes seem to centre particularly on effeminacy, coercion, the seduction of minors, and the participation of citizens in passive acts. Bullough (1976) suggests that social disapproval was considerably less for boys, women, and slaves who participated in such acts because they were excluded from the Roman power structure and played a passive role in society. Julius Caesar was himself subjected to ridicule after rumours that he had played a passive role in his sexual relations with Nicomedes, King of Bithynia. His soldiers are said to have chanted 'Caesar conquered Gaul, Nicomedes Caesar'. This concern about passivity evidently declined later in the Empire, probably because some emperors themselves favoured sexual passivity.

According to Greenberg and Bystryn (1982), similar social disapproval of male passivity has prevailed in Germanic tribes, feudal Japan, the Moslem Middle East, and New Guinea. Indeed, they cite Middle Assyrian laws against the spreading of false rumour concerning passive homosexuality and suggest that similar attitudes were held in ancient Egypt and Mesopotamia. Incidentally, they indicate that, in the sexual area, the Middle Assyrian code specifically forbids only the homosexual rape of a neighbour.

A liberal attitude to sex, lack of distinction between homosexual and heterosexual behaviour, and preoccupation with status and passivity changed with the rise of Christianity and its accompanying asceticism. The new asceticism was generally hostile to all forms of sexual pleasure. Indeed, by the fifth century the Roman physician Caellius Aurelianus grouped passivity and opposite-gender identification together as a mental disorder. However, these generalizations have to be qualified. Boswell (1980) quotes at least six prosecutions for homosexual crimes in the First century AD, but he claims the common theme was again the abuse of free-born Roman citizens of either gender. In each the defendant was either an adult who had tried to seduce an minor who was the son of a Roman citizen, or a civil or military officer who had attempted to force a subordinate to gratify him. It appears that the crucial issues determining public retribution were the social status and rights of the accused versus those of the victim.

The most celebrated case quoted was that of a young soldier in the army of Gaius Marius who killed a tribune when the latter coerced or forced him into sexual relations. The issues at the trial were the rights of a subordinate to slay a superior who abused him. While the case became famous, it was not until several centuries later that Roman law unequivocally upheld the right of a male to defend himself against sexual violence by killing his attacker.

In another case quoted by Boswell (1980), a young Roman citizen enslaved for debt was severely beaten when he did not submit to sexual advances. The case caused sufficient concern for it to become a major issue leading to the convening of the Senate to discuss it. But the central issue seems to have been the propriety of enslaving a Roman citizen for debt. Had the victim been a slave in the ordinary sense, it appears most improbable that any case would have been brought.

Greece

In ancient Greece (Greenberg and Bystryn 1982), homosexual behaviour was seen as an expression of love of a high order and occurred in a variety of forms. These authors describe its existence in the Thebean army as an expression of comradeship in arms among noble warriors that was so effective as to render the army undefeatable. However, in the Classical period in Greece, the situation appears to have been somewhat different and usually involved a greater discrepancy in age between participants, thereby increasing the coercive aspect. Indeed, this coercive pattern of behaviour is not uncommon in highly patriarchal societies where there is personal domination by a male over all the other members of the family. In this type of society: 'all sexual relations are conceptualized in terms of power involving superordination and subordination'; just as men may dominate women in heterosexual relationships, so one man dominates another man or boy in a homosexual relationship (Greenberg and Bystryn 1982). In a patriarchal society, this type of relationship might be considered demeaning for the subordinate partner if there was no disparity in age or power, but not for the controlling partner, who simply confirms his dominant status. Thus an adolescent male could be dominated by older men without incurring stigma because this was the natural order of things in a partriarchal society.

Early and medieval Christian

The rise of Christianity within Roman society and later throughout Europe was accompanied by more austere attitudes to sex and later by specific condemnation of homosexual behaviour. However, many sects existed in the early church, as they did in the Middle Ages. Each had a variety of doctrines, later defined as heretical, among which some gnostics regarded women as common property and sexual intercourse as a sacred religious mystery. While licentious behaviours occurred, their attitude to heterosexual rape is unclear.

The edict of Otto the First, promulgated in Rome in AD 966 but based on an earlier Roman edict of AD 390, prescribed burning and strangulation for some homosexual acts, probably for homosexual rape. Ecclesiastical denunciation of homosexual behaviour reappeared strongly in the eleventh century, with particularly disapproval being

Cultural and historical aspects of male sexual assault 109

directed at clergy who practised sodomy. Later these sentiments were incorporated into secular legislation, prescribing the death penalty for intercourse 'outside the fit vessel' – behaviour later loosely called sodomy of both men and women (Greenberg and Bystrun 1982). In 1108, the Council of Westminster condemned men wearing their hair long, and leading churchmen preached against the effeminacy of the royal court, but it appears that homosexuality was deplored only incidentally. The deepest preoccupation was with men dressing and acting like women and showing passive behaviour.

By the late eleventh and early twelfth century a letter attributed to Alexius Conmenus to Count Robert of Flanders, imploring his aid against the Turks, exemplifies these attitudes (Boswell 1980) and the prevailing Christian view of Islam: '... Oh incomparable Count, great Defender of the Faith, it is my desire to bring to your attention the extent to which the most Holy Empire of the Christian Greeks is fiercely beset every day by the Pincinnatti and Turks... for they circumcise Christian boys and youths over the Baptismal fonts of Christian churches and spill the blood of circumcision right into the Baptismal fonts and compel them to urinate over them... When they capture noble women and their daughters, they abuse them sexually in turn like animals... but what next, we pass on to worse yet. They have degraded by sodomising men of every age and rank, boys, adolescents, young men, old men, nobles, servants and what is worse and more wicked clerics and monks and even – alas and for shame! Something which, from the beginning of time has never been spoken or heard of – Bishops. They have already killed one Bishop with this nefarious sin.'

In 1440, Gilles de Rais (Allan 1969), a companion of Joan of Arc, is said to have killed 200 boys, for sexual pleasure, over about an 8–year period before his execution. He said, 'I found a Latin book on the lives and customs of the Roman Caesars by the learned historian called Suitonius. The said book was ornamented by pictures very well painted in which were seen the manner of these pagan emperors and I read in this fine history how Tiberius, Caracula and other Caesars sported with children and took singular pleasure in martyring them, upon which I desired to imitate the said Caesars and the same evening I began to do so following the pictures in the book... The children killed were thrown into a vat at the foot of a tower... except for a few handsome heads I kept as relics.'

Greenberg and Bystrun (1982) suggest that homosexual practices were common amongst medieval knights in Europe and amongst similar groups in Japanese and Islamic cultures. These relationships appeared to stem from extremely close attachments, often of many years, leading to a homosexual relationship, which, they suggest, was based on power.

Islam

Bullough (1976) summarizes the Islamic position as follows: 'Islam has put itself into a dilemma. It allows polygamy and concubinage which, in the light of known sex ratios often means that there are not enough women for every man to have a wife. In addition, it strictly segregated men and women making it difficult to establish positive relationships between the sexes... All male social life was spent in the company of men, yet sex generally was regarded as good and pleasurable. With such contrary attitudes homosexuality became a viable alternative for many and even heterosexuals regarded it as a possibility.'

In the Koran, homosexual behaviour is specifically condemned: Surrah Al-Shu' are (The Poets) No.XXVI, verses 165, 166, 'Come you unto the males of the people and leave those whom Allah has created for you to be your wives? Nay you are a long grieving people.' In the Hadji – the 'Saying of The Prophets' – any anal penetration is prohibited but it seems that force used in this context is not specifically mentioned. Thus, the position in Islam, in so far as one can generalize, is that homosexual behaviour is condemned while, at the same time, women are considered inferior. It is known, however, that despite this ambivalence, homosexual acts do occur. In Islamic law, consensual and non-consensual homosexual acts remain punishable by death, as does adulterous behaviour by women.

China

Bullough (1976) describes extensive homosexual practices in the Tang dynasty, AD 618–907, among male actors (women were forbidden to appear on the stage). Homosexuality was considered the norm among them and the junior actors were expected to be at the sexual service of the more senior. The juniors were said to be abject participants in highly ritualized acts of anal dilation before participating in these homosexual acts. Later, during the Ming dynasty, there is said to have been a more liberal attitude to sexual matters but this in turn was succeeded by a particularly repressive period (Qu'ing, 1644–1912). The relevant penal code, the Qu'ing Code, had a section on sexual violations. While earlier codes were maintained, the act of rape was redefined so that proof of violation became extremely difficult. This attitude to rape of a female probably influenced attitudes to male rape. The victim had to provide evidence that she had struggled with her assailant throughout the entire act. Evidence had to include witnesses, bruises, and torn clothing. Moreover, if the victim submitted, however late the submission was, the act was considered 'illicit intercourse by mutual consent' and for this the victim was punished. This emphasis on corroboration of a rape allegation and the equation of submission with willingness is echoed in current European legal practice.

Politically, according to Ng (1987), this change could be understood as part of an effort to discourage charges of rape against Manchu soldiers who had recently pacified the area and a concern that the new regime should be accepted by the populace. The code included detailed legal dispensations for soldiers committing the following sexual acts: abduction and gang rape, rape and murder, rape aggravated by a non-fatal injury, rape of an adult male, rape of a boy under 10 (years), rape of a boy between 10 and 12 (years) with and without consent, and consensual sodomy.

This precise categorization may have been used as a means of controlling not only rape but homosexuality as well, which Ng suggests was not at all uncommon in that period. This control, he suggests, was congruent with the new period of sexual repression consequent upon a period of absolutism in government, with a resultant push for intellectual and institutional uniformity. He quotes a specific case in which, during the trial, it was discovered that the victim had had prior homosexual relations with another man and this revelation significantly affected the outcome of the trial. Although the rapist was convicted his defence was mitigated, from strangulation after the Assizes to 100 blows and banishment. The victim was found guilty of previous sodomy and given the statutory sentence of 100 blows with a heavy bamboo, plus wearing the cangue for one month. The aggressor paid a higher punishment price than he would had he been found guilty of sexually assaulting an unchaste woman, and the victim received a greater punishment than would a hypothetical female victim. This suggests that while men were held to be of greater value than women, they were at the same time held more responsible for their actions.

Eighteenth century Europe

Ray (1988) quotes eighteenth century authors as estimating the number of Parisian 'sodomites' in 1725 variously as 20 000 and 40 000. This probably refers to homosexuals. In this context he describes the trial of Benjamin Deschauffours, who was arrested for kidnap, raping, and selling boys to French and foreign aristocrats after he killed a boy. He was burned at the stake. Contemporary accounts were said to be unanimous in emphasizing the contribution of sodomy to the sentence. A play of the period, *L'Ombre de Deschauffours*, analyses the trial in political terms as an attack by heterosexuals on homosexuals.

In his memoirs of the period, Lenoir (the Lieutenant-General of Police from 1775 to 1785) is quoted as describing the position from the point of view of the police: 'At the time of sentence, Paris numbered ... more than 20 000 individuals who offered examples of the vice for which Deschauffours mounted the scaffold. What was wanted was a political chastisement ... the executioner worked to this end. In the long run pederasty can only be a vice of gentlemen.'

Although the acts for which Deschauffours was charged were murder, rape, and procurement, and the punishment, typical of the period, was burning at the stake, the political motive was control of homosexual behaviour. It seems probable that similar attitudes prevailed in Britain (Davenport-Hines 1990).

The writings of de Sade include lurid accounts of sadistic sexual assaults on men and boys. Perhaps his most famous book, *120 Days of Sodom* (de Sade 1954, reprinted), was regarded at the time with horror and disbelief, although more recent literary critiques by Camus, Simone de Beauvoir, and Raymond Barthe have interpreted his writings as a creative attempt to explore previously unimaginable areas of sexual phantasy through literature (de Beauvoir 1955).

An ethological approach

An ethological approach to male sexual assault in which analogous behaviour in non-human species, particularly primates, is studied would seem intrinsically useful and relevant. This type of approach may allow not only identification of analogous behaviour but also of behaviour with evolutionary roots in common with that seen in man. It might even be possible to show the development of the behaviour across species from an earlier evolutionary form to that seen in man.

Wickler (1967) suggests that behaviour evolved for one function, in this case sexual, can be used for other purposes such as social signals. However, the concept of intent, a central notion in human sexual assault, cannot be applied to an animal. We are forced to examine analogous behaviour rather than cognition. Males mounting males is not an uncommon phenomenon among mammals. It occurs in marsupials (Buchmann 1990) and ungulates (Ewer 1968), although frequently the mounting is incomplete or from the side, rather than over the back. In the rhesus monkey (*Macaca mulatta*) mounting of males by males may occur in a pattern behaviourally identical to that seen in heterosexual mounting by these animals. Although this may include anal penetration, it is usually of briefer duration than the heterosexual act and ejaculation does not normally occur. Carpenter (1964) describes the act occurring in juvenile play in groups that are exclusively male, and occasionally in heterosexual situations. Wickler (1967) describes similar behaviour in baboons, without specifying intramission, and interprets this as part of a dominant-submissive relationship, the mounting male being the more dominant.

The gestures adopted by both participants are those of heterosexual mounting, the submissive male assuming feminine postures. Indeed, females may mount females in a similar way, using respectively male postures in the mounter and female postures in the mounted.

Squirrel monkeys (*Saimiri sciureus*) frequently use penile display (Ploog et al. 1963) as part of an agonistic interaction between males during the mating period. This, too, can be understood as a dominant act outside the heterosexual context but these displays do not progress to mounting or intramission. Eibl-Eibesfeldt (1971) provides innumerable ethological, ethnological, and archaeological examples of male sexual displays used for agonistic purposes.

There have been objections to using concepts of dominance and hierarchy to understand this behaviour. Rowell (1974) prefers the concept of 'social role', seeing the animal performing a variety of behaviours, the primary purpose of which is to allow it to function effectively in the social context in which it finds itself. As she suggests, hierarchies may be a product of captivity or augmented by captivity. This concept is equally relevant for our purpose. In heterosexual groups, according to Carpenter (1964), the new member seems to gain the tolerance of the dominant male by using submissive postures, while amongst the dominant males the submissive postures may best be understood as an indication of affinity.

Some would argue that these gestures are performed 'willingly' by the subordinate and in this sense are more analogous to homosexual acts than to rape. However, the context under which these acts occur are essentially non-sexual, so that a more accurate way of regarding the behaviour is that sexual gestures are here being used as expressions, respectively, of dominance and appeasement of potential aggression.

There is considerable variation in the frequency with which sexual gestures are seen in various primate species, being unusual in the gorilla and orang-utan and common in the chimpanzee. This frequency is probably reflected in the degree to which sexual gestures are used for social functions. However, there does seem to be a general propensity for sexual gestures to be linked with social dominance/submissive social situations in most primates.

Conclusions

Although historical and ethological data are meagre in this area, those that do exist appear to indicate the existence of male sexual assault over a wide variety of species, cultures and times. The socio-biological material, in particular, suggests that these acts are more likely a manifestation of power relationships than of sexual ones. The historical data are largely in accord with this view, although it is clear that major differences across cultures can occur as a consequence of different social attitudes. These historical and socio-biological findings appear to be in accord with modern thinking about homosexual rape, particularly in institutions where the acts are best seen as acts of dominance. From a psychodynamic point of view, similar proposals have been made (Fried 1970). However, to regard these acts as either a sexual or a power one may be a false dichotomy. In man the act appear to serve the function of both power and sexual gratification simultaneously, and the question then becomes how much of one and how much of the other in any particular interaction.

References

Allan, C. (1969). *A textbook of psychosexual disorders*. Oxford University of Press.
Boswell, J. (1980). *Christianity, social tolerance and homosexuality*. University of Chicago Press.
Breiner, S. J. (1990). *Slaughter of the innocents: Child abuse through the ages and today*. Plenum Press, New York.
Bullough, V. L. (1976). *Sexual variance in society in history*. Wiley, New York.
Carpenter, C. R. (1942). Sexual behaviour of free ranging rhesus monkeys (*Macaca mulatta*) II. *Journal of comparative psychology*, **33**, (1), 243–62.
Davenport-Hines, R. (1990). *Sex, death and punishment. Attitudes to sex and sexuality in Britain since the Renaissance*. Collins, London.
de Beauvoir, S. (1955). *The Marquis de Sade. Faut-il-brûler Sade*. Gallinast, Paris.
de Sade, D. A. F. (edn. 1954) *120 Days of Sodom*, (trans. P. Excavini), Olympia Press, Paris.
Eibesfeldt, I. (*op. cit.*).
Eibl-Eibesfeldt, I. (1971). *Love and hate*. Methuen, London.
Ewer, R. F. (1968). *Ethology of mammals*, pp. 143–86. Logos Press, London.
Fried, E. (1970). *Active and passive: The crucial psychological dimension*. Brunner/Mazel, New York.
Greenberg, D. F. And Bystryn, M. H. (1982). Christian intolerance of homosexuality. *American Journal of Sociology*. **88**, 515–47.
Ng, V. M. (1987). Ideology and sexuality: Rape laws in Qing (Ch'ing) China. *Journal of Asian Studies*, **46**, 57–70.

Ploog, D.W., Blitz, J., and Ploog, F. (1963). Studies on the social and sexual and behaviour of the squirrel monkey (*Saimiri sciureus*) *Folio Primatologidy* **1**, 29–66.
Ray, M. (1988). Police and sodomy in 18th century Paris: from sin to disorder. *Journal of Homosexuality*, **16**, 129–46.
Rowell, T. E. (1974). The concept of social dominance. *Behavioural Biology*, **11**, 131–54.
Scacco, A. M. (1981). *Male rape, a case book of sexual aggression*, A.M.S. Studios in Sociology No. 15 A.M.S. Press, New York.
Sykes, G. M. (1958). *N.Y. Society of Captives – a study of a maximum security prison*. Wiley, New York.
Tiger, L. (1969). *Men in groups*. Quoted in Eibl-Eibesfeldt, I. (*op.cit*).
Wickler, W. (1967). *Socio-sexual signals and their intra-specific imitation among primates in private ethology*, (ed. D. Morris), pp. 69–147. Weidenfeld & Nicholson, London.

8
Male victims of sexual assault – legal issues
Zsuzsanna Adler

Introduction

The complexity of the law on sexual offences is perhaps best illustrated by the vast range of prohibited behaviours. The report of a Howard League working party (1985) on unlawful sex lists no less than 46 different offences including rape, buggery, incest, gross indecency between men, living on earnings of prostitution, and indecent exposure. These offences differ along a number of important dimensions such as consent, sex of the perpetrator and of the victim, the nature of the activities involved, and so on. As Walmsley and White (1979) have helpfully summarized, a person commits a sexual offence if the behaviour in question fulfils at least one of the following conditions:

- it takes place without the consent of the other party;
- it takes place with a person under the age of consent;
- it is in itself prohibited by law;
- it is homosexual behaviour not committed in private; or
- it is homosexual behaviour by a male himself under the age of consent.

In addition to being complex, the law in this area lacks internal consistency. It makes no visible distinction between different levels of anticipated harm: for example, whether it is the protection of youngsters from premature sexual experiences, the protection of people from violence, or indeed the prohibition of acts seen as undesirable in themselves that is at issue, the law deals with the crime in much the same way. Maximum penalties for different offences do not always reflect any coherent scale either of perceived wrongdoing or of harm to the victim. The law also discriminates between males and females in this area, both as victims and offenders. For example, the age of consent for heterosexual and homosexual acts is different, and while homosexual acts between men are subject to legal controls, similar behaviours between women are not.

Whatever view one takes of the desirability of such discrepancies, their existence is not altogether surprising as legislation tends to reflect the norms, standards, and sexual morality that prevailed at the time when particular laws were passed or key precedents were set. As Honore (1978) observed, 'no moral theory has a monopoly of the law, which is a patchwork of statutes and cases put together at different times and by people of varying outlook. Some bits reflect a moral viewpoint, others a different one.'

The purpose of this chapter is to review the current law on sexual offences committed against men; to compare that in its scope and operation to the law on sexual offences committed against women; to explore the fundamental principles that underlie differences between the two; and finally to make some observations about the degree to which the law in practice protects men from sexual victimization. It will be argued that the attention that female sexual victimization and its treatment by the criminal justice system have received in recent years has served to highlight the most blatant injustices in that area. No parallel developments have taken place for male victims of similar crimes; indeed, it is arguable that adult male victims of sexual assault may, in some respects, have even more formidable obstacles to overcome in reporting and substantiating sexual victimization than do their female counterparts.

Law on sexual offences against men

The main indictable offences to be discussed here are buggery, incest, indecent assault on a male, and gross indecency between males. What these offences have in common is that their victims may be males and, in the case of indecent assault on a male and gross indecency, they may only be males. However, there is also much to distinguish them, in terms of the sex of the perpetrator, the nature of the behaviours covered, and last but not least, the issue of consent. These factors will now be discussed in turn.

Let us first take the sex of the perpetrator. For buggery and gross indecency, the 'perpetrator' is invariably male. Indecent assault may be committed by males or females, but one study found that only 1 per cent of those convicted of indecent assault on a male were females (Walmsley and White 1979). Incest with males, on the other hand, is necessarily committed by females.

As far as victims are concerned, according to the strict letter of the law, some of these crimes are necessarily homosexual while others are, or may be, heterosexual. For example, one study shows that approximately 10 per cent of offences of buggery recorded by the police are heterosexual (Walmsley and White 1979). That research further indicates that around 20 per cent of sexual offences known to the police in the years before their study had been homosexual offences, although roughly 40 per cent of persons convicted of a sexual crime had been convicted of a homosexual offence. They argue that this discrepancy between recorded offences and convictions is largely due to the nature and processing of two offences, namely indecency between males and unlawful sexual intercourse with a girl under 16 years of age. The former has a high clear-up rate, a high prosecution rate, and usually involves two offenders per offence. Consequently, it has a high conviction rate per offence known. Unlawful sexual intercourse with a girl under 16, by contrast, has a very low prosecution rate, and hence, a similarly low conviction rate per offence known. Thus, the distinction between homosexual and heterosexual offences appears to be important in the operation of the law in this field.

Let us now turn to the nature of behaviours covered by the above offences. Buggery involves sexual intercourse *per anum*. Incest is defined as sexual intercourse with a person of the opposite sex where there exists a close blood relationship, specified in the legislation, between the parties involved. The law has yet to spell out the meaning of indecency for the purposes of defining offences such as gross indecency and indecent assault: its interpretation is largely left to the discretion of those administering and enforcing the law. It appears to include any behaviour that may be described as sexual but falls short of intercourse. The following extract from a case (*R. v. Willis* [1974] 60 Cr. App. R(S) 149) of indecent assault against a male illustrates the inherent subjectivity in the interpretation of what actually constitutes an offence of indecent assault, as well as of the perceived seriousness of different behaviours by one judge:

It must be remembered that in these cases it is not the label of indecent assault which is important but the nature of the act. In many cases it amounts to no more than putting a hand on or under clothing in the region of the testicles or buttocks. Such cases are not serious. In some the assault may take the form of a revolting act of fellatio, which is as bad as buggery, maybe even more so.

The most interesting aspect of the law here is that offences such as buggery and incest are prohibited behaviours in themselves. Whether the parties are willing participants in the behaviour involved or not is immaterial, with one exception: homosexual acts in private no longer constitute an offence provided the parties have attained the age of consent and do, in fact, consent. For an indecent assault to take place, by contrast, there must be a 'hostile act, a threatening gesture or a threat to use violence' (Howard League Working Party 1985).

Consent is defined by the *Oxford English Dictionary* as 'voluntary agreement or acquiescence in what another proposes or desires'. The concept is important. As far as those sexual offences that can be committed against males are concerned, the law takes three distinct approaches. The first is applied to offences such as incest, where the law does not recognize consent as a defence under any circumstances. The second approach is to recognize consent as a defence, provided the victim is above a certain age. The law makes certain assumptions about the age at which people can give informed consent to sexual behaviour, and sees as one of its functions the protection of those considered too immature to make considered choices in these matters. Furthermore, as mentioned above, different assumptions are made about a person's ability to make such decisions according to whether the sexual behaviour contemplated is homosexual or heterosexual: the age of consent for the two is 21 and 16 years, respectively. Thirdly, for those above the age of consent, the law treats the issue of consent as a true variable.

The net effect of existing provisions is that the question of consent, insofar as potential male victims of sexual offences are concerned, only arises in an extremely limited number of situations. The vast majority of unlawful sexual behaviours involving male 'victims' are, in fact, unlawful whether the parties are consenting or not. The only exception is indecent assault against a male.

This raises some fundamental questions as to whom or what the law is trying to protect and control in this area. What are the features of sexual behaviours that make them into a crime?

A review of the above offences suggest that one, and it might be argued, limited, purpose of the law is to protect adult males from sexual victimization, that is, from involvement in acts to which they do not consent. Another of its aims is to protect boys from homosexual and heterosexual acts to which they do not consent, or to which they are deemed too young to give informed consent. This is the case for indecent assault on a male, and may be case for buggery and incest.

However, there seems to be another major intention behind the law governing sexual offences involving males, and that is to control certain consensual behaviours. Such behaviours include three different kinds of acts. First, those that are seen as somehow unnatural and undesirable in themselves (e.g. buggery); secondly, sexual intercourse between people linked by blood ties (e.g. incest); and third, homosexual behaviour outside certain well-defined boundaries (e.g. gross indecency). Although it is difficult to decide whether primacy should be given to the protective or control elements of the law, it may be argued that much of the legislation concerning sexual crimes involving males is designed to regulate consensual homosexual behaviours rather than to protect males from sexual victimization.

How do these imperatives compare with the law governing sexual offences committed against females?

Law on sexual offences against females

The main indictable sexual offences against females are rape, incest, buggery, indecent assault, and unlawful sexual intercourse. These are offences against females in the sense that their victims may be female, as in incest and buggery, or that they are necessarily female, as in the case of the other offences.

As far as the perpetrators of these offences are concerned, they are almost exclusively male, with the exception of indecent assault, which can also be committed by a female. In practice, this is a rare occurrence and there are few such convictions. In addition, women have occasionally been convicted of aiding and abetting rape, although they cannot be charged with the full offence. This immediately highlights a major difference between sexual crimes against men and women: as we saw, the former tend to involve homosexual behaviour, while the latter in the main involve heterosexual acts.

The prohibited acts, notwithstanding differences in the sex of the perpetrator, overlap with those prohibited where males are concerned. Buggery and incest, for example, are identically defined whatever the victim's sex. Indecent assault is just as vaguely defined for females as it is for males. However, unlawful sexual intercourse and rape both involve a narrow definition of sexual intercourse as penetration of the vagina by the penis. An additional complication with rape is that, until 1991, it was not an offence when committed by a husband on his wife, a situation that has been much criticized in recent years (Mitra 1979; Williams 1984; Ginsburg and Lerner 1989).

The notion of consent is crucial in the definition of sexual offences against females, although as with males, its presence or absence does not operate identically for different offences. As far as incest and buggery are concerned, whether the female consents in fact is immaterial to the question of whether or not a crime had been committed. The essence of unlawful behaviour is, as with men, the commission of an act that is in itself deemed to be undesirable. It may be thought of as more undesirable in the absence of consent, but remains a serious crime in any event.

As far as unlawful sexual intercourse with a girl under 13 or 16 years of age is concerned, and for indecent assault on a girl under the age of consent, the law treats the victims of such crimes as incapable of giving informed consent – thus, absence of consent is assumed, rather like for young male victims of indecent assault. The essence of what is against the law here is the sexual exploitation of young persons.

With indecent assault against females over 16, and rape, the issue of consent is markedly different. The essence of those crimes is not the behaviour involved, or the vulnerability of its victims, but the fact that the acts complained of were committed in the absence of the woman's consent.

With rape, the definition of consent has given rise to much debate over the last 15 years or so. Questions have arisen as to whether there need to be present violence or threats of violence for lack of consent to be present; how important the defendant's interpretation of the absence or presence of consent is; and how far consent to intercourse with a particular individual can be inferred from the woman's sexual history and background (Report of the Advisory Group on the Law of Rape 1975; Coote and Gill 1983; Adler 1987; Temkin 1987).

Until the recent ruling by the House of Lords (R v R [Rape: marital exemption], *The Times*, 24th October 1991), the legal protection of women from rape did not extend to married women where rape by their husbands was concerned. In such cases, although lack of consent may have been only too patently obvious, the behaviour was not previously against the law, which was based on the principle that marriage implies consent to intercourse at all times. This was the reverse of some of the situations discussed above, where an act remains a crime even where both parties consent. Since the recent decision, the general rule is that where one party does not consent to a sexual activity, this is invariably a crime. With children, there is an additional factor, which is an assumption in law that below a certain age they cannot consent to sexual acts.

Thus, the main thrust of the law on sexual offences against females is to protect victims from unwanted sexual assaults. A further implied aim of the law in this area is the protection of the freedom of sexual choice of women. The exception to that is unlawful sexual intercourse, which aims to protect girls from premature sexual acts, whether they are willing participants or not. Nevertheless, the overall thinking here is in marked contrast with the law on sexual offences against males, where the prime intention seems to be the control of homosexual behaviour.

The operation of the law

The above discussion centres on the law as it appears in its legislative form. However, this should be distinguished from the law is it is routinely practised and enforced. The disparity between the two has been commented upon with regard to legislation on rape and allied offences (Adler 1982), and in a different context, Smart (1981) has made the following observation:

> Apparently progressive legislation can be rendered virtually useless if the courts and other agencies fail to operate it as legislators intended, whilst the fact that the legislation exists at all sustains a liberal and egalitarian front to actual legal practices.

This is not to imply that the law is particularly liberal or egalitarian where sexual crimes are concerned; the above discussion shows rather the opposite. However, the point is that the law is not a unified structure and its operation can vary according to its different levels of jurisdiction. It can also be dependent on the activities of the police and other legal personnel who are in a position to operate discretion. Finally and equally importantly, the operation of the law is largely dependent on the willingness of the public to report crimes that have occurred.

In order to make some assessment as to how effectively the law protects males and females from sexual victimization, this section will review the criminal statistics relating to indecent assault, which is one sexual offence that in principle operates in the same way whether the victim is male or female.

It has been observed that 'official statistics have many limitations . . . and the literature abounds with caveats regarding their interpretation. Inevitably, they suffer from inaccuracies of recording, they are dependent on what is reported, and indeed they are dependent on what is classified as criminal' (Walmsley and White 1979).

Those limitations are fully recognized. Nevertheless, for the purpose of comparing the way in which the criminal justice system processes recorded crimes of indecent assault against males and females, they may be argued to constitute a useful starting point.

The letter of the law has already been discussed above. One matter that is noteworthy, however, is the relatively recent equalization of the maximum penalties for indecent assault against males and females. Before 1985, the maximum penalty for indecent assault on a female aged 13 years or over was two years' imprisonment. The equivalent for a female under the age of 13 was five years. Where the victim of indecent assault was male, however, irrespective of his age, the maximum penalty was 10 years. Whatever the historical origins of this discrepancy, it undoubtedly reflected some notion of the perceived relative seriousness of such assaults against males and females. In its report on sexual offences, the Criminal Law Revision Committee (1984) recommended that indecent assault should be punishable by a maximum of 10 years' imprisonment, irrespective of gender, and this recommendation has since been adopted by Parliament in the Sexual Offences Act 1985.

In 1988, criminal statistics show 2512 offences of indecent assault against males recorded by the police, compared to 14 112 similar offences against females (Criminal Statistics 1988). A comparison of the relevant figures over the last 10 years indicates that offences against females have consistently outnumbered those against males in this sort of ratio. There have been roughly five times as many crimes against females than males over this period. However, the ratio has increased from 4.6 in 1984 to 5.6 in 1988. This is clearly associated with the marked increase in absolute terms of offences against females, particularly since the mid-1980s. In the 10 years between 1978 and 1988, known offences against males had increased by 2 per cent. The comparable figure for offences against females is 19 per cent.

Are there also differences in the processing of offences of indecent assault against males and females? The only data bearing on this in criminal statistics relate to findings of guilt for the two crimes. The proportion of those convicted of indecent assault against males has been marginally higher every year between 1978 and 1988 than of those convicted of crimes against females. For males, the range of conviction rates has been between 28 and 36 per cent, compared to 26 and 30 per cent for females. However, conviction rates for males have steadily if marginally fallen since 1983, and are now lower than they were in 1978. For crimes against females, the figures have been much more steady over the same years.

It is admittedly very difficult to make meaningful interpretations of these figures. What is clear, however, is that the recording and processing of offences of indecent assault against males and females are far from being identical. There are marked sex differences in the numbers of recorded crimes of indecent assault offences. Crimes against females are a great deal more prevalent and growing faster than crimes against males. Furthermore, there is a hint that perpetrators of offences against males are pursued more vigorously than perpetrators of crimes against females. The clear-up rates in 1988 were 90 per cent for crimes against males and 66 per cent for crimes against females. Similarly, the conviction rates of offenders against males have been consistently, if only marginally, higher than of offenders against females in the last 10 years. There is also some evidence that this trend has been evening out since the mid-1980s.

Unfortunately, criminal statistics do not tell us anything beyond the classification of offences and findings of guilt. In particular, they do not contain information about victims, perpetrators, or any measure of the seriousness of various crimes of indecent assault. A Home Office study sheds some light on this area (Walmsley and White 1979). During the year of study, the authors found that 88 per cent of the indecent assaults on males leading to a conviction involved victims under 16, while the comparable figure for females was 70 per cent. In the absence of similar figures for all recorded indecent assaults, it is impossible to draw any conclusions regarding prosecution policy. What is clear, though, is that a significantly higher proportion of convictions for indecent assault against males involve youngsters than is the case with indecent assault against females. This may go some way towards explaining a more marked tendency to clear up and achieve convictions in those offences, insofar as crimes against children are in general perceived as more serious than crimes against adults.

Discussion

The foregoing has shown that the letter of the law on sexual offences differs in several important ways according to the sex of the victim. In particular, offences against males, often equated with homosexual offences, are sharply distinguished in law from offences against females, which tend to involve heterosexual acts. As a Howard League report (1985) argues:

Boys involved sexually with adults present particular problems because the offenders are nearly always male and the incidents therefore homosexual. Heterosexual contact between boys and older females are rare, or at any rate rarely give rise to complaints.

Commentators have noted that the law often ascribes greater seriousness to homosexual offences. One illustration of this is the fact that consent in homosexual crimes is less of an issue than it is for heterosexual ones: it is often the behaviour in itself, rather than the fact that it is practised without the other party's consent, that is deemed unlawful.

Walmsley's study of sentencing practice, however, indicates that the courts are not always inclined to follow this view. For instance, before 1985, while there was a disparity in the maximum penalties for indecent assault on males and females, the length of sentence imposed for the two offences was quite similar. Against that, it should be noted that custodial rates for indecent assault against males were somewhat higher than they were for females (20 per cent versus 12 per cent). This was the same across different age bands for victims as well as offenders.

Another aspect of the operation of the law in this area concerns the numbers of sexual crimes recorded against males and females. As we have seen in the case of indecent assault, there appears to be a very significant difference in rates of male and female victimization.

Whether this reflects the true incidence of such crimes in the community is difficult to ascertain. A retrospective study on the prevalence of sexual victimization among children showed that 10 per cent of boys and 12 per cent of girls had been sexually abused before the age of 16 years (Baker 1985). The figures on recorded crimes show a different picture. However, it is interesting that a large majority of male victims of reported indecent assault are, in fact, children. It is reasonable to hypothesize that even with boy victims, the reporting of sexual victimization is low compared to girls. It has been argued (McMullen 1990) that:

... the prevalance rates recorded in the literature concerning the abuse of boys are probably underrepresentative of actual abuse occurrence rates. Currently they average out at approximately 2.5 female disclosures for every male, but there are special factors relating to the kind of research so far conducted which can lean more to the identification of female oriented abuse rather than male.

But what happens to adult males when they become victims of sexual assaults? Do the recorded figures reflect the incidence of male sexual victimization in the community? Is this very much lower than female victimization, as criminal statistics indicate? Alternatively, is it possible that the law in this area, both in theory and practice, acts as a deterrent to reporting these offences?

Any attempt to interpret the data available on male victims of sexual crime and the ability of the law to protect them adequately must be, to some extent, speculative. We have seen that the law on sexual crimes involving male victims is primarily concerned with the control of homosexual behaviour. We have also seen that the recorded rate of male sexual victimization is very much lower than its equivalent for females. Furthermore, a very large proportion of male victims of indecent assault are boys under the age of 16.

In general the criminological literature, where it concerns itself at all with male victims of sexual crime, takes a fairly complacent view of the situation. For example, West (1987) sees homosexual offences as matters of immorality or indecency with the young rather than true assaults, arguing that such crimes rarely involve violence and are often consensual. Furthermore he argues that prosecutions for gross indecency between adult males mostly concern conduct in public, often around public lavatories, and are in the nature of things consensual, even when one or other participant pleads otherwise in an effort to escape conviction.'

Drawing on data from the study by Walmsley and White (1979), West asserts that the indecent assaults on males that lead to a criminal conviction almost invariably involve young boys, and that 'many of these cases concerned fully consensual activity'. The Howard League report (1985), in a similar vein, states that compared to girls who are victims of indecent assault 'boys are more likely to be active collaborators rather than unwilling recipients of adult sexual attentions, if only because reported incidents occur when they are older and in situations in which they could escape'.

The evidence upon which such statements are based is questionable, and appears to come from one Home Office study, which examined sexual offences against boys and girls under the age of consent. It made no attempt to assess consent in sexual behaviour in children under the age of 10, but above that age, classified behaviour as consensual 'where the documentary sources gave clear evidence to that effect' (Walmsley and White 1979). According to those rather vague criteria, 80 per cent of homosexual crimes and 34 per cent of heterosexual crimes were found to have been consensual in the year of study. Having said that, Walmsley and White acknowledged that one of the functions of the law is to protect those considered too immature to make informed choices in these matters, and that their classification as 'consensual' of behaviour involving a person under age does not imply that their consent is always as valid as that of an older person.

West (1987) also asserts that the only context where males are at any significant risk of sexual assault is when they are confined to prisons or similar institutions. In general, criminologists have argued that males are at risk of sexual victimization to a lesser extent than females; that when they do become 'victims' this is only in a technical sense because they tend to consent to the acts involved; and that the only real exception to this is the victimization of males in prisons and similar institutions. The main impact of observations such as the above is to place male sexual victimization in a context that is far removed from the everyday context of most males in our society, and to depict the phenomenon as extremely rare and out of the ordinary.

Such a view of male sexual victimization is reminiscent of the general perception of female sexual victimization, and rape in particular, some 15 years ago. At that time, it was thought of as a relatively rare phenomenon, involving predominantly young women and men unknown to them, and usually characterized by the use of considerable force. Since then it has been recognized that such a definition is only applicable to a small minority of rape cases. However, the above description was indeed typical of those rapes that then came to the attention of the authorities. Since then, research has shown that rape is far more common than was previously thought; that women are most likely to be assaulted in their own homes or in the assailant's home; that most victims are known to their assailants; and that relatively few victims suffer serious physical injury.

Some American researchers have in fact begun to describe a pattern of male sexual victimization outside institutions (Groth and Burgess 1980; Kaufman et al. 1980; Goyer and Eddleman 1984). These studies suggest that male sexual victimization is a neglected area of study, often conceived 'as an aberration of prison life, a form of vicarious rape against women, or as a violent outgrowth of the homosexual subculture' (Kaufman et al. 1980). Similarly, in the United Kingdom, McMullen (1990) argues that it is a mistake to assume that male sexual victimization is a rare event simply because so few instances are either reported or known about in legal, medical, or academic circles.

Among the many developments in the study of female sexual victimization over the last 15 years, four are particularly noteworthy insofar as they may help our understanding of current perceptions of the prevalence and treatment of male sexual victimization. First, a pattern of typical reactions to sexual assault has been identified and described. Secondly, the institutional and societal responses to victims, and the way in which these affect post-assault reactions have been documented. Thirdly, it has been found that the motives for sexual assault are complex, and that they have more to do with anger, power, and a desire to humiliate than with sex. Finally, the propensity of many victims to remain silent about their assault has been recognized.

Female victims of sexual crime are liable afterwards to experience fear, anxiety, shock, embarrassment, shame, and irrational feelings of guilt. These feelings are often reinforced and amplified by the negative reactions of their social network, particularly where the assault does not fit the stereotype of the classical stranger attack, or where the victim is criticized for failing to offer enough resistance. The fear of encountering such reactions may be, and often is, strong enough for victims to decide to remain silent about their assault. Even after years of concentrated effort by various agencies of the criminal justice system to encourage more women to come forward, and despite some progress in this area, crimes like rape remain very much underreported.

Male victims of such crimes are likely to experience similar feelings. However, they may also have additional fears connected with ideas of homosexual contamination and loss of masculinity. McMullen (1990) argues that 'the stigmatisation of male rape victims by their attackers is a distressingly prevalent consequence of an attack'. Where the victim is male, any claim that he consented projects on to him a homosexual identity. Where the victim is homosexual, this can lead to considerable feelings of guilt, which tend to act as a deterrent to reporting. Where the victim is heterosexual, the very fear of being thought a homosexual may well stop him from reporting. In fact, the reasons for not reporting for male victims are much the same as they are for female victims, and include shock, embarrassment, fear, self-blame, and a high degree of stigma.

We have seen above that the law governing sexual crimes against males is primarily concerned with the control of homosexual behaviour, while the law on sexual crimes against females aims to regulate non-consensual heterosexual behaviour. This distinction may contribute considerably to the under-reporting of sexual crimes by males. In describing male rapes in a prison setting, Sagarin (1976) commented as follows:

That the act itself was homosexual is indisputable from the viewpoint of a behavioural scientist. But this is not quite so clear to the participants. They were concerned not with the definition of the act, but with the labelling of the actors, including themselves.

This raises a fundamental question about the validity of labelling sexual crimes against males as 'homosexual'. As far as the definition of the actors is concerned, this may not be the case. In fact, some go further and assert that 'male rape is rarely, if ever, a homosexual problem'.

Various studies have shown that it is misleading to conclude from the sexual content of crimes against males that the offender is homosexual and that the crime is therefore homosexually motivated. As with crimes against women, motivating factors appear to have little to do with sexual gratification, and a great deal to do with power and aggression. Victims for both kinds of crimes tend to be selected for their vulnerability and their limited ability to defend themselves. Indeed, it has been argued that some rapists are gender-blind, and are likely to assault both males and females. Several researchers have shown that only a relatively small proportion of male offenders against other males are, in fact, homosexual (McGeorge 1964; Rossman 1979).

If the offender tends not to be homosexual, what of the victim? It is particularly difficult in this area to obtain reliable data, particularly as it is likely that homosexual victims are less likely to report than others. Such research as is available shows that there exist both homosexual and heterosexual male victims of sexual crime, and that the crime has a different impact on the two groups, at least in some respects.

The stigma of homosexuality that is so clearly attached to the law governing sexual crimes against males must act as a major deterrent to reporting such crimes. It is arguable that the major intention of the law regarding sex crimes against males should be the protection of males from sexual attacks and victimization. Much has been achieved since the mid-1970s to ensure that the law acts in this way where female victims are concerned. The time has come to transfer some of the lessons learnt during that time to male victims. In particular, for progress to be made in this area, the homosexual label that is attached to such crimes should be modified; and the difficulties experienced by men in reporting such crimes to the authorities must be acknowledged and taken into account. There appears to be no legitimate reason for the law to discriminate between male and female victims in this area. Yet discriminate it does: for example, the anonymity given to women reporting rape and indecent assault is not given to male victims of sexual crime. Similarly, the thinking behind the need to legislate differs where male and female victims are concerned. As the Howard League's Report (1985) argued:

... the predominant concern about heterosexual molestation of girls is that they may find the experience off-putting, and thereby become anxious, frigid or even lesbian, the concern for boys involved homosexually with older men is that they may find the experience attractive and be seduced into homosexuality.

Empirical research to date indicates that fears of a child being somehow converted to a homosexual orientation as a result of such victimization are unfounded. The evidence is that all victims, irrespective of age or gender, find the experience of sexual assault frightening, traumatic, and stigmatizing, although individual variations do exist. From the above review, it would appear that the law on sexual crime makes a better job of acknowledging this for female victims than it does for males. It seems likely that the very low rates of reporting by male victims do not reflect low prevalence rates of such crimes, but rather the law's unwillingness to concern itself more positively with the protection of male children and adults from unwanted sexual attacks.

References

Adler, Z. (1982). Rape: the intention of Parliament and the practice of the courts.
Modern Law Review, 45, 664–75.

Adler, Z. (1987). *Rape on trial*. Routledge & Kegan Paul, London.
Baker, A. W. (1985). Child sexual abuse: a study of prevalence in Great Britain. *Child Abuse and Neglect*, **9**, 457–67.
Coote, A. and Gill, T. (1983). *The rape controversy*. National Council for Civil Liberties, London.
Criminal Law Revision Committee (1984). *Fifteenth report – sexual offences*. HMSO, London.
Criminal Statistics, England and Wales (1988). HMSO, London.
Ginsburg, E. and Lerner, S. (1989). *Sexual violence against women – a guide to the criminal law*. Rights of Women, London.
Goyer, P. E. and Eddleman, H. C. (1984). Same sex rape of non-incarcerated men. *American Journal of Psychiatry*, **141**, 576–9.
Groth, N. and Burgess, A. W. (1980). Male rape – offenders and victims. *American Journal of Psychiatry*, **137**, 806–10.
Honore, T. (1978). *Sex law*. Duckworth, London.
Howard League Working Party (1985). *Unlawful sex*. Waterlow Publishers, London.
Kaufman, A., Divasto, P. Jackson, R., Voorhees, H., and Christy, J. (1980). Male rape victims: non-institutionalised assault. *American Journal of Psychiatry*, **137**,
McGeorge, J. (1964). Sexual assaults on children, *Medicine Science and the Law*, **4**, 245–53.
McMullen. R. J. (1990). *Male rape – breaking the silence on the last taboo*. GMP Publishers, London.
Mitra, C. (1979). . . . For she has no right or power to refuse her consent. *Criminal Law Review*, 558–65.
Report of the Advisory Group on the Law of Rape (1975). HMSO, London.
Rossman, P. (1979). *Sexual experience between men and boys*. Maurice Temple Smith, London.
Sagarin, E. (1976). Prison homosexuality and its effect on post-prison sexual behaviour. *Psychiatry*, **39**, 245–57.
Smart, C. (1981). Law and the control of women's sexuality. In *Controlling women – the normal and the deviant*, (ed. B. Hutter and G. Williams). Croom Helm, London.
Temkin, J. (1987). *Rape and the legal process*. Sweet and Maxwell, London
Walmsley, R. and White, K. (1979). *Sexual offences, consent and sentencing*, Home Office Research Study No. 54. HMSO, London.
West, D. J. (1987). *Sexual crimes and confrontations*. Gower, Aldershot.
Williams, J. (1984). Marital rape – time for reform. *New Law Journal*, **134**, 26–8.

9
Treatment for male victims of rape
Gillian C. Mezey

Rape is a life-threatening assault that precipitates the victim into a state of chaos. The majority of victims recover from the crises and adapt in the same way as individuals adapt to bereavement and other life-threatening trauma. However, the event is rarely forgotten and frequently results in changes to the individual's attitudes in behaviour towards himself, strangers, friends, and family (Mezey and King 1989; McMullen 1990). Little has been written about the response of men after serious sexual assault or the treatment that should be offered. Rape is a crime predominantly directed against women and therefore an underlying assumption of treatment provision is that rape counsellors must necessarily be female. Rape crisis centres tend to be informal and driven by feminist philosophy and men are viewed as the aggressors rather than the victims. There is only one organization in Great Britain claiming to address itself specifically to the needs of male victims of sexual assault.

With increasing reports of sexual assaults, more men are featuring amongst the victims. There is so far little evidence to suggest that men react in a fundamentally different way from women after such assault. They express the same sense of disbelief, fear, humiliation, and rage that such a thing could have happened and that they could have allowed it to happen. Male victims experience rape as a profound attack on their sexuality, an assault driven by anger and aggression that violates their physical and psychological boundaries. The effects of rape are analogous to post-traumatic stress disorder (PTSD), a pattern of reactions consisting of psychological, cognitive, behavioural, and physiological effects (American Psychiatric Association 1987), and although the majority of victims recover, a number of studies suggest significant longer-term psychiatric problems, including PTSD in a large number of victims. A community study by Kilpatrick *et al.* (1987) found that 16.5 per cent of rape victims still had PTSD an average of 17 years after a sexual assault. The impact of sexual abuse in childhood may be even more devastating: reactions in adult survivors include raised levels of depression and anxiety (Burnam et al. 1988; Mullen et al. 1988), suicidal ideation (Briere and Zaidi 1989; Herman and Hirschman 1981), substance abuse (Burnam et al. 1988; Briere and Zaidi 1989; Shearer et al. 1990) and Axis-2 disorders (American Psychiatric Association 1987), particularly borderline personality disorder (Briere and Zaidi 1989; Herman et al. 1989; Shearer et al. 1990).

Gender differences in response to trauma

Gender differences in response to disaster have not been systematically studies. However, a number of investigators have suggested that women show more anxiety than men and are more ready to accept and seek out help after a disaster (e.g. Milgram et al. 1988). Male victims of a community disaster were described as being more belligerent and as abusing alcohol to a great extent than did female victims (Green et al. 1990. Male victims of child sexual abuse have a tendency to externalize feelings of anger whereas anger in women is more likely to be self-directed and present as depression, low self-esteem, and deliberate self-harm (Carmen et al. 1984). One of the few studies on adult male victims of sexual assault concluded that male victims are more ready to acknowledge and express anger (Groth and Burgess 1980), although more recent work has failed to replicate these findings (Mezey and King 1989). The apparent differences in male – female response to disaster may be a reflection of sex-role socialization, whereby emotionality in men is interpreted as a sigh of weakness and vulnerability, evidence that they are less than 'real men'. In spite of these differences, most evidence suggests that male victims of sexual assault experience very similar responses to those of females. It is therefore likely that counselling techniques that benefit women should be equally helpful in treating men.

There is a remarkable lack of information on the subject of treating male victims of sexual assault and no empirical data. Reports on work with male victims of sexual assault have been largely anecdotal, and this work tends to have targetted children and adolescents rather than adult victims. Burgess and Holmstrom (1979) reported on work with six male victims aged from 3 to 21 years. They confirm that the needs and responses of males are very similar to those of female victims except in that the overwhelming concern of the male victims, social network was whether the attack would make the victim gay.

The following section will review results of empirical studies, the underlying assumptions of treatment, and issues arising in the course of treating victims of sexual assault. There remains uncertainty about several fundamental questions regarding treatment: first, what form should any intervention take; secondly, which group of victims should be targetted and at what point after the assault; thirdly, whether 'treatment' in the early stages can prevent long-term progression to more chronic disability; finally, how should services be delivered and what resources should be allocated to enable this to be carried out efficiently?

Who provides treatment?

In Great Britain there are few treatment options for female rape victims and virtually none for men who are sexually assaulted. This compares unfavourably with the United States, where the first hospital-based, rape-crisis intervention programme was established in the early 1970s (McCombie et al. 1976) and most hospitals have training programmes for staff dealing with the effects and treatment of rape-induced PTSD. Treatment is comprehensive and involves physical examination, psychological counselling, and practical assistance – for example, finding alternate accommodation, providing transport for the victim, and ensuring their physical safety.

Very few of these centres, however, claim to see more than a handful of adult male victims each year, and many centres acknowledge their lack of skill and expertise in dealing with men who report rape. In the states of America where the legal definition of rape has become gender-neutral, increasing numbers of men are prepared to report and seek out treatment. Several investigators have advocated the incorporation of similar sexual-assault centres into major hospitals in Great Britain (Duddle 1985; Mezey 1987). Providing treatment for victims, particularly victims of crime, does not appear to be a priority in health care in Great Britain and, to date, there is virtually no comprehensive, multidisciplinary, treatment provision, little co-ordinated research, and an absence of state provision outside the funding of voluntary organizations. Treatment, in particular long-term psychological counselling, arises out of philanthropy rather than being recognized as the victim's 'right'.

One criticism of many programmes of service delivery for rape victims is that they place an emphasis on their crisis but pay little attention to longer-term psychological problems. Second is that services are often established without any reference to the stated needs of the victims themselves or reference to existing empirical research data (Maguire 1985). The recently published *Victims Charter* (Home Office 1990) emphasizes the government's commitment to improving the experience of the victim within the criminal justice system and supports the work of victim support schemes, yet fails to acknowledge the limitations of volunteer organizations (Corbett and Maguire 1988). No mention is made in the document of the need for more professional involvement, perhaps because of the inevitable and potentially profound resource implications. The reluctance of the medical profession to see rape as a psychiatric problem and their tendency to share many of the stereotypes and prejudices surrounding sexual assault has been noted by a number of commentators (Notman and Nadelson 1976; Magure and Corbett 1987).

The role of volunteer organizations

In Great Britain, most treatment is offered in the community by volunteers: rape crisis centres and the National Association of Victims' Support Schemes (NAVSS) being the two major organizations. The NAVSS is a volunteer organization providing treatment in the form of crisis intervention to victims of crime, including serious sexual assault. In addition to offering short-term psychological and emotional help, volunteers provide practical assistance such as rehousing, accompanying the victim to hospital for medical checks, applying for compensation, and assisting the victim through the court process (Corbett and Hobdell 1988).

The NAVSS at present only take on victims who are referred by the police. Under-reporting is a significant problem in rape and may be even more extensive amongst male victims. Therefore, unless victim support schemes broaden their referral base, they have little to offer the majority of male victims of sexual assault. In their philosophy and practice the schemes are essentially non-feminist and non-political, unlike rape crisis centres, which were originally inspired through feminist activity and which regard rape as a manifestation of societal power inequality and misogyny rather than the product of individual pathology.

The London Rape Crisis Centre was the first centre established in Great Britain in 1974 and is representative of this view (London Rape Crisis Centre 1984). The Centre does not employ male volunteers and does not offer counselling or advice to men who attempt to contact them. The Centre provides telephone counselling but victims are discouraged from speaking to the same counsellor on follow-up to prevent them becoming dependent on the Centre or on any one volunteer. The disadvantage of this practice is that it disrupts continuity and may prevent the woman from establishing a relationship of trust with any one person. In contrast, victim support schemes have less of a political emphasis in their interpretation of rape, they operate outreach both at the initial interview and at follow-up, and they offer an immediacy, frequency, and duration of support not provided by rape crisis centres. Unlike rape crisis centres, they encourage collaboration with the victim's social network, including police and other professional workers. Unfortunately, partly because of the limited amount of resources available, and partly because of a virtually unbridgeable philosophical and political divide, there is considerable competition rather than collaboration between these two agencies and other services for sexual assault victims.

Adequate training for counsellors is essential whatever their status. This must include the availability of supervision and support for the counsellor, and the possibility of professional help for victims who develop chronic problems and are in need of more specialized help. Without adequate training and supervision, therapists may over-identify with the victims and find themselves becoming over-protective, or alternatively, defend against their own sense of vulnerability by becoming angry with the victim, criticizing him for having allowed it to happen or responding to the story with disbelief. Although there is a danger that victims may be adversely affected by untrained, insensitive or inept individuals, the situation in Great Britain is that, without volunteers, victims of sexual assault and other serious crimes would be unlikely to get any assistance (Maguire and Corbett 1987). There has been some attempt to evaluate the work of victim support schemes, although most of this has been based on the subjective reports of women who have been seen (MaGuire and Corbett 1987). They lack standardized measures of assessment and the numbers that have been seen are too small to draw any firm conclusion. However, rape victims may find the notion of volunteer assistance less stigmatizing than psychiatric treatment provided in the setting of a psychiatric hospital (King and Webb 1981). Rape crisis centres in Great Britain tend not evaluate their work, largely because they regard research as antithetical to the core notion of reinvesting the rape victim with a sense of control and of 'victims' being essentially healthy as opposed to 'mentally ill' (Anna, T. 1988).

The form of treatment

General issues

Research into the efficacy of treatment interventions for 'rape trauma syndrome' is mainly descriptive (Holmes and St. Lawrence 1983). Veronen and Kilpatrick (1983) looked at the prognosis for untreated rape victims at one year and found that only 17 to 25 per cent were symptom-free. Burgess and Holmstrom (1979) carried out a 5–year telephone follow-up with over 100 rape victims and found that 37 per cent had recovered within months, for 37 per cent it had taken years, and 26 per cent still considered themselves not recovered.

It is clear that, with or without therapeutic intervention, the majority of victims of sexual assault show a rapid resolution of symptoms over the first four to six weeks (Kilpatrick *et al.* 1979 *a*, and *b*). One study demonstrated that, whereas 97 per cent of rape victims met criteria for PTSD within a week of the assault, only 46 per cent met the criteria at 9–week assessment (Rothbaum and Foa 1988). There is therefore little point in offering treatment in the immediate recovery phase, as improvement attributed to crisis intervention could equally be due to spontaneous automatic recovery.

Burgess and Holmstrom (1974*a*), amongst others, stress the importance of allowing the victim to ventilate their feelings soon after an attack. By this they mean allowing the victim to describe what has occurred – his cognitive, physical, and emotional response and conflicts arising from the experience – without being criticized or judged by others. This simple process of ventilation may facilitate communication with the victim's family and friends, and mobilize his social network, enabling them to provide more effective support. There is some empirical evidence that simple expression of a traumatic event, without any formal treatment being offered, is in itself an effective anxiety-reducing exercise (Pennebaker and Susman 1988). Several researchers have noted that simply participating in research may be therapeutic: victims seen several times, for the purpose of assessment only, show a trend towards lower levels of depression and better social adjustment than those subjects who are assessed less frequently (Kilpatrick *et al.* 1979*a*; Resick *et al.* 1981; Atkeson *et al.* 1982.

Whatever the form of treatment given, there are specific core issues that are recognized as being fundamental to the development of post-traumatic pathology after sexual assault, and that have to be addressed in treatment. It has been advocated that the word 'survivor' should be used rather than 'victim', underlying the expectation that recovery will occur. However, it is generally unhelpful to have a political ideology imposed upon a victim of a serious violent sexual assault at a stage when they are often at their most vulnerable and suggestible. Although rape is predominantly a violent assault, the sexual component of rape should not be down-played at the expense of the violent nature of the act. It is precisely because sex is involved that victims of rape, more than any other crime, experience disbelief, blame, and isolation. It is not uncommon for the victim to respond physiologically to the sexual part of the act, while being repelled and terrified by the aggression and brutality. Whereas victims of non-sexual violence can expect sympathy and unconditional support from friends and family, victims of sexual assault, whatever their gender, experience far more ambivalence and are often aware of others' curiosity as being intrusive and voyeuristic rather than helpful. Victims of rape need to be given anticipatory guidance and to recognize that, for the most part, their distress is essentially normal and will resolve within a period of a few months.

Other central issues that must be dealt with in treatment are the profound sense of loss (Whiston 1981) and the rape victim's tendency to feel guilty and to blame themselves (Libow and Doty 1979). Victims of sexual assault appear to be unwilling, or unable, to express anger towards the rapist. This inability is most usefully explained in terms of attribution theory, which allows the victim to gain a sense of control over the future and an illusion of safety by attributing responsibility for what has happened to themselves rather than the perpetrator. Victims of rape frequently dread re-experiencing the rape in therapy and may defend themselves against feelings of overwhelming anxiety by denial and avoidance. There is little evidence that the gender of therapist is important in influencing the victim's rate of recovery: the requirement of many rape crisis centres that only women deal with female rape victims is based more on political rather than scientific considerations. A more crucial issue is the individual's sensitivity, training, and attitude towards the offence (Bassuk and Apsler 1983).

Crisis intervention

Crisis intervention (Lindemann 1944) has been advocated as the treatment of choice for survivors of a rape attack, based on the pioneering work of Burgess and Holmstrom 1974b). Their original description has subsequently been criticized for its lack of empiricism and the failure to detail their methods of treatment or to evaluate its efficacy in any way. There has been no research that has demonstrated the superiority of crisis intervention over simple support, assessment, or no treatment, either in terms of alleviating acute symptoms in the rape victim or in preventing longer-term psychiatirc disability (Kilpatrick and Veronen 1983). The underlying principles of crisis intervention are that it is short-term, focused work, generally consisting of 10–12 sessions, it takes place over a period of a few months, and the therapist takes an active role in initiating follow-up. The individual is assumed to have had a normal personality before the assault and should return to that pre-existing normal personality within a short period of time. A feature common to victims of all forms of major trauma is their reluctance to continue in treatment and their unwillingness to maintain follow-up (Baisden and Quarantelli 1978; Lindy et al. 1981). Veronen and Kilpatrick (1983) found that less than half of rape victims assessed as being in need of help took up offers of treatment. This was attributed to the therapy becoming, by its association with the rape, an aversive stimulus through second-order conditioning. In another study, less than a quarter of victims completed a 14–hour treatment programme (Frank and Duffy-Stewart 1983). Victims who seem to be most in need of help and most damaged also appear to be least able or willing to seek out and make use of treatment (Mezey and Taylor 1988). For chronic victims, a form of learned helplessness may be a reason for failing to seek out help (Seligman 1975); the high attrition rate is also related to victims' extensive use of denial and avoidance, and the unstable lifestyle created by the rape (Ellis et al. 1981). Another reason for failure to continue in treatment is the tendency of the victim's social network to form a protective 'trauma membrane' around the victim, which may prevent would-be counsellors from making contact (Lindy et al. 1981).

Specific treatment interventions

Psychodynamic psychotherapy

Although there are some reports of short-term dynamic group therapy (Cryer and Beutler 1980; Perl et al. 1985), and the individual psychodynamic psychotherapy (Werner 1952; Factor 1954), they generally lack control (non-treatment) groups, fail to include any systematic evaluation of their efficacy, and concern very small numbers of subjects. However, psychodynamic theory is certainly of some value in terms of understanding, interpreting, and dealing with the many complex issues that arise with this group of clients (Notman and Nadleson 1976; Rose 1986. Sprei and Goodwin (1983) and Yassen and Glass (1984) describe time-limited group therapy based on feminist philosophy that deals with issues of self-esteem, trust, power, loss of control, guilt, mourning, and anger. Each claimed benefit from the treatment although this was based on the subjective reports of victims alone.

Cognitive behaviour treatments

The most promising treatment strategies appear to be those that provide the victim with specific mechanisms and alternative responses to manage their anxiety (Holmes and St. Lawrence 1983). In practice, these involve a combination of education, cognitive, and behavioural techniques.

Cognitive restructuring is aimed at reframing the victimization experience in a more positive and constructive light, and in challenging the victims' misinterpretations and distorted assumptions that may hinder their recovery. Social-skills training includes relaxation and other coping skills, for example, anger management, problem-solving, and assertiveness training. The aim of systematic desensitization is to decrease the anxiety associated with aspects of the rape by the use of imaginal exposure and relaxation; it is effective in reducing intrusive memories, flashbacks, and nightmares associated with the rape.

Wolff (1977) successfully used systematic desensitization and negative practice to treat a rape victim's fears, seven years post-assault. By contrast, Becker and Abel (1981) were unable to show any positive effect of systematic desensitization and negative practice in the treatment of rape-related phobias in two rape victims. Forman (1980) has described improvement in a rape victim's obsessive thoughts on using cognitive restructuring and thought-stopping.

Cognitive behavioural treatments have been used effectively to treat certain chronic symptoms after sexual assault, in particular fear and anxiety, sexual dysfunction, and depression.

Frank and Duffy-Stewart (1983) treated 17 victims of sexual assault over 14 sessions with cognitive restructuring and systematic desensitization. The cognitive therapy included self-monitoring of activities, and identification and correction of distorted attitudes, as originally described by Beck (1976). They reported an improvement in ratings of fear, anxiety, depression, and social adjustment, results being similar with the two forms of treatment. However, all subjects had been recently assaulted and no control groups were used.

Prolonged imagination and *in vivo* exposure to fear cues have been used to treat the effects of sexual assault and abuse (Rychtarik *et al.* 1984) but their use with adult victims of rape have been criticized on ethical grounds (Kilpatrick and Best 1984).

A brief behavioural intervention programme (BBIP; Kilpatrick and Veronen 1983) for recent victims of rape consisted of 2–3 sessions, usually lasting 4–6 hours, and included relaxation training, an explanation of the origin of self-blame and rape fears, and taught specific coping skills to the victims. There has been no evidence, however, that the BBIP was any more effective than simple assessment alone.

'Stress inoculation training' appears to be a promising cognitive behavioural approach for the treatment of victims who remain fearful more than three months after being raped (Kilpatrick *et al.* 1982). An important part of the treatment package is an initial educational phase that explains the origin of rape-related fears through classical conditioning theory and the manifestation of these through physical, emotional, and cognitive channels. Techniques of stress inoculation training include relaxation and deep breathing, role-playing and covert modelling, thought-stopping and guided self-dialogue. Veronen and Kilpatrick (1983) treated 15 female victims of rape who, after three months, were still experiencing fear and anxiety to specific rape-related cues, with 20 hours of stress inoculation training. No control groups were used; subjects showed a significant improvement on measure of rape-related fear, anxiety, phobic anxiety, tension, and depression.

Rothbaum and Foa (1988) have presented preliminary findings for comparison of the efficacy in rape victims of stress inoculation techniques, prolonged imaginal exposure, and supportive counselling with a 'no treatment' group. All women received nine 90–minute individual sessions at twice-weekly intervals and were assessed pre- and post-treatment and at three months, using a number of validated instruments and a structured interview. Although numbers in each group were small, both stress inoculation techniques and prolonged imaginal exposure were found to be superior to supportive counselling, and a 'waiting-list' group, in reducing symptoms of PTSD.

The results of both these studies are limited, however, in that the subjects were highly selected in having circumscribed target phobias and raised levels of anxiety as their main problems. It is uncertain whether similar behavioural – cognitive techniques would be equally effective in treating the more diffuse and complex disturbances arising as a result of chronic victimization or childhood sexual abuse.

Sexual dysfunction has been described as a long-term consequence of sexual assault in both male (Mezey and King 1989) and female victims (Burgess and Holmstom 1974a, b; Becker et al. 1979; Feldman-Summers et al. 1979; Nadelson et al. 1982; Becker and Skinner 1983). These reports suggest that it is primarily individuals' sexual enjoyment that is affected as a result of the assault rather than the actual frequency of sexual activity. Behavioural approaches to the treatment of sexual dysfunction have been shown to be effective (Becker and Abel 1981; Becker and Skinner 1983), particularly when the victim's partner is included rather than individual therapy. Incest victims show less improvement than adult victims of recent sexual assault.

More work is needed to determine the type of support required by male victims of sexual assault. Until male rape is recognized as a legal entity, the extent of the problem will be underestimated and the victims will continue to be marginalized. Many treatment agencies only offer help within the framework of criminal justice. Therefore, by defining male rape out of existence, the law effectively denies the victims protection and assistance.

References

American Psychiatric Association (1987). *Diagnostic and Statistical Manual of Mental Disorders*, (3rd edn, revised) APA, Washington DC.

Anna, T. (1988). Feminist responses to sexual abuse: The work of the Birmingham Rape Crisis Centre. In *Victims of crime: a new deal*, (ed. M. Maguire and J. Pointing. Open University, Milton Keynes.

Atkeson, B. M. Calhoun, K. S. Resick, P., and Ellis, E. M. (1982). Victims of rape: repeated assessment of depressive symptoms. *Journal of Consulting and Clinical Psychology*, 50, 96–102.

Baisden, B. and Quarentelli, E. (1978). The delivery of mental health services in community disasters: an outline of research findings. *Journal of Community Psychology* 9, 195–203.

Bassuk, E. and Apsler, R. 1983). Are there sex biases in rape counselling? *American Journal of Psychiatry*, 140, 305–8.

Beck, A. T. (1976). Cognitive therapy and the emotional disorders. N.Y. International Universities Press.

Becker, J. V. and Abel, G. G. (1981). Behavioural treatment of victims of sexual assault. In *Handbook of clinical behaviour therapy*, (ed. S. M. Turner, K. S. Calhoun, and H. E. Adams). Wiley, New York.

Becker, J. V., Abel, G. G. and Skinner, L. J. (1979). The impact of a sexual assault on the victims sexual life *Victimology*, 5, 229–35.

Briere, J. and Zaidi, L. Y. (1989). Sexual abuse histories and sequelai in female psychiatric emergency room patients. American Journal of Psychiatry, 146, 1606–6.

Burgess, A. W. and Holmstrom, L. L. (1974a). Rape trauma syndrome. *American Journal of Psychiatry*, 131, 981–6.

Burgess, A. W. and Holmstrom, L. L. (1974b). *Rape: victims of crisis*. Brady, Bowie MD.

Burgess, A. W. and Holmstrom, L. L. (1979). *Rape: crisis and recovery*. Brady, Bowie MD.

Burnam, M. A. *et al.* (1988). Sexual assault and mental disorders in a community population. *Journal of Consulting and Clinical Psychology*, 56, 843–50.

Carmen, E. H., Rieker, P. P., and Mills. T. (1984). Victims of violence and psychiatric illness. American Journal of Psychiatry, 141, 374–83.

Corbett, C. and Hobdell, K. (1988). Volunteer-based services to rape victims: some recent developments. In *Victims of crime: a new deal*, (ed. J. Pointing and M. Magure). Open University, Milton Keynes.

Corbett, C. and Maguire, M. (1988). The value and limitations of victim support schemes. In *Victims of crime: a new deal*, (ed. M. Maguire and J. Pointing). Open University, Milton Keynes.

Cryer, L. and Beutler, L. (1980). Group therapy: an alternative treatment approach for rape victims. *Journal of Sex and Marital Therapy*, 6, 40–6.

Duddle, M. (1985). The need for sexual assault centres in the United Kingdom. *British Medical Journal*, 290, 771–3.

Ellis, E. M., Atkeson, B. M., and Calhoun, K. S. (1981). An assessment of long-term reactions to rape. *Journal of Abnormal Psychology*, **90**, 263–6.

Factor, M. (1954). A woman's psychological reaction to attempted rape. *Psychoanalytic Quarterly*, **28**, 243–4.

Feldman-Summers, S., Gordon, P., and Meagher, J. (1979). The impact of rape on sexual satisfaction. *Journal of Abnormal Psychology*, **38**, 101–5.

Forman, B. D. (1980). Cognitive modification of obsessive thinking in a rape victim: a preliminary study. *Psychological Reports*, **47**, 819–22.

Frank, E. and Duffy-Stewart, B. D. 1983). Treating depression in victims of rape. *Clinical Psychologist*, **36**, 95–8.

Green, B. L. et al. (1990). Buffalo Creek survivors in the second decade: stability of stress symptoms. *American Journal of Orthopsychiatry* **60**, 43–54.

Groth, A. N. and Burgess, A. W. (1980). Male rape: offenders and victims. *American Journal of Psychiatry*, **137**, 806–10.

Herman, J. and Hirschman, L. (1981). Families at risk for father – daughter incest. *American Journal of Psychiatry*, **138**, 967–70.

Holmes, M. R. and St. Lawrence, J. A. (1983). Treatment of rape-induced trauma: proposed behavioural conceptualization and review of the literature. *Clinical Psychology Review*, **3**, 417–33.

Home Office (1990). *Victims charter: a statement of the rights of victims of crime*. Home Office, London.

Kilpatrick, D. G. and Best, C. L. (1984). Some cautionary remarks on treating sexual assault victims with Implosin. *Behaviour Therapy*, **15**, 421–3.

Kilpatrick, D. G. and Veronen, L. J. (1983). Treatment for rape-related problems: crisis intervention is not enough. In *Crisis intervention*, Vol. IV, *Community Clinical Psychology Series*, (2nd edn.), (ed. L. H. Cohen, W. L. Claiborn, and G. A. Specter). Human Science Press, New York.

Kilpatrick, D. G., Veronen, L. J., and Resick, P. A. (1979a). The aftermath of rape: recent empirical findings. *American Journal of Orthopsychiatry*, **49**, 658–60.

Kilpatrick, D. G., Veronen, L. J., and Resick, P. A. (1979b). Assessment of the aftermath of rape: changing patterns of fear. *Journal of Behavioural Assessment*, **1**, 133–48.

Kilpatrick, D. G. Veronen, L. J. and Resick, R. A. (1982). Psychological sequelae to rape assessment and treatment strategies. In *Behavioural medicine: assessment and treatment strategies*, (ed. D. M. Doleys, R. L. Meredith, and A. R. Ciminero). Plenum Press, New York.

King, M. E. and Webb, C. (1981). Rape crisis centres: progress and problems. *Journal of Social Issues*, **37**, 93–104.

Libow, J. A. and Doty, D. W. (1979). An exploratory approach to self blame and self derogation in rape victims. *American Journal of Orthopsychiatry*, **49**, 670–9.

Lindemann, E. (1944). Symptomatology and management of acute grief. *American Journal of Psychiatry*, **101**, 141–8.

Lindy, J. D., Grace, M. C. and Green, B. L. (1981). Survivors: outreach to a reluctant population. *American Journal of Orthopsychiatry*, **51**, 468–78.

London Rape Crisis Centre (1984). *Sexual violence: the reality for women*, Handbook Series. Women's Press, London.
McCombie, S. L. Bassuk, E., Savitz, R., and Pell, S. (1976). Development of a medical rape crisis intervention programme. *American Journal of Psychiatry*, **133**, 418–21.
McMullen, R. J. (1990). *Male rape: breaking the silence on the last taboo*. GMP Publishers, London.
Maguire, M. (1985). Victims needs and victims services: indications from research. *Victimology*, **10**, 539–59.
Maguire, M. and Corbett, C. (1987). *The effects of crime and the work of victims support schemes*. Gower, Aldershot.
Mezey, G. (1967). Hospital based rape crisis programmes: what can the American experience teach us. *Bulletin of the Royal College of Psychiatrists*, **11**, 49–51.
Mezey, G. and King, M. (1989). The effects of sexual assault on men: a survey of 22 victims. *Psychological Medicine*, **19**, 205–9.
Mezey, G. and Taylor, P. J. (1988). Psychological reactions of women who have been raped. *British Journal of Psychiatry*, **152**, 330–9.
Milgram, N. A., Toubiana, Y. H., Klingman, A., Raviv, A., and Goldstein, I. (1988). Situational exposure and personal loss in children's acute and chronic stress reactions to a school bus disaster. *Journal of Traumatic Stress*, **1**, 339–52.
Mullen, P. E., Romans-Clarkson, S. E., Walton, V. A., and Herbison, G. P. (1988). Impact of sexual and physical abuse on women's mental health. *The Lancet*, **i**, 841–5.
Nadelson, C., Notman, M. Zackson, H., and Gornick, J. (1982). A follow-up study of rape victims. *American Journal of Psychiatry*, **139**, 10.
Notman, M. and Nadelson, C. (1976). The rape victim: psychodynamic considerations. *American Journal of Psychiatry*, **133**, 408–12.
Pennebaker, J. W. and Susman, J. R. 1988). Disclosure of traumas and psychosomatic processes. *Social Sciences and Medicine*, **26**, 327–32.
Perl, M., Westin, A. B., and Peterson, L. G. (1985). The female rape survivor: time-limited group therapy with female – male cotherapists. *Journal of Psychosomatic Obstetrics and Gynecology*, **4**, 197–205.
Resick, P., Calhoun, K., Atkeson, B., and Ellis. E. (1981). Social adjustment in victims of sexual assault. *Consulting and Clinical Psychology*, **5**, 705–12.
Rose, D. (1986). Worse than death: psychodynamics and the need for psychotherapy. *American Journal of Psychiatry*, **143**, 817–24.
Rothbaum, B. and Foa, E. (1988). *Treatments of post-traumatic stress disorder in rape victims*. Paper presented at World Congress of Behaviour Therapy Conference, Edinburgh, Scotland (Sept).
Rychtarik, R. G., Silverman, W. K., Van Landingham, W. P., and Prue, D. M. (1984). Treatment of an incest victim with implosive therapy: a case study. *Behaviour Therapy*, **15**, 410–20.
Seligman, M. E. P. (1975). *Helplessness: on depression, development and death*. Freeman, San Francisco.
Shearer, S. L., Peters, C. P., Quaytman, M. D., and Odgen, R. L. (1990). Frequency and correlates of childhood sexual and physical abuse histories

in adult female borderline patients. *American Journal of Psychiatry*, **147**, 214–17.

Sprei, J. and Goodwin, R. A. (1983). Group treatment for sexual assault survivors. *Journal for Specialists in Group Work*, **8**, 39–46.

Veronen, L. J. Kilpatrick, D. G. (1983). Stress management for rape victims. In *Stress reduction and prevention*, (ed. D. Meichenbaum and M. E. Jarenko, pp. 341–74. Plenum Press, New York.

Werner, A. (1972). Rape: interruption of the therapeutic process by external stress. *Psychotherapy: Theory, Research and Practice*, **9**, 349–51.

Whiston, A. K. (1981). Counselling sexual assault victims: a loss model. *Personnel and Guidance Journal* (Feb), 363–6.

Wolff, R. (1977). Systematic desensitisation and negative practice to alter the after-effects of a rape attempt. *Journal of Behaviour Therapy and Experimental Psychiatry*, **8**, 423–5.

Yassen, J. and Glass, L. (1984). Sexual assault survivors groups: a feminist practice perspective. *Social Work*, **29**, 252–7.

Index

acting out 44–5
adolescents
　homosexual relationships 17, 19
　sexual abuse by 36–8, 45–50, 57
　sexual abuse of 27–58
age
　of consent 19, 119, 120–1
　onset of sexual abuse 34
agents provocateurs 20
aggressive behaviour, sexually abused boys 43–4
AIDS 18, 22, 82
　see also HIV infection
alcohol use 4, 7, 53
anal abuse, children and adolescents 39
anal intercourse, *see* buggery
anger
　counsellors towards rape victims 101
　female rape victims 101–2
　male co-survivors of rape victims 90–1, 93, 94
animal behaviour 112–14
armed forces 16, 72
assailants (perpetrators)
　children and adolescents as 36–8, 57
　factors promoting 45–7
　identification of risk 47–50
　community male rape 4
　gender 1–2, 117, 120
　prison rape 68, 71
　sexual abuse 17, 34–9
　sexual orientation 1–2, 4, 9, 17
attribution theory 137
avoidance reactions 76, 77–8, 81, 83

behaviour disorders, sexually abused boys 44
blackmail, homosexuals 18, 21
brief behavioural intervention programme (BBIP) 139
buggery (anal intercourse) 117, 118, 120
　consensual 19, 20
　forced 1, 5

Child Behaviour Checklist (CBCL) 44
children
　consent to sexual behaviour 126
　identification of perpetrator risk 47–50
　physical abuse 38, 50
　sexual abuse, *see* sexual abuse
　sexual abuse by 36–8, 45–50, 57
China 110–11
Christianity, early and medieval 107, 108–10
Cleveland Inquiry into child abuse (1988) 39
cognitive behaviour treatments 138–40
cognitive responses, male co-survivors of rape 89–92
cognitive restructuring 138–9
communication problems, rape survivors and their partners 93–4
community, male sexual assault 1–11, 127
consent
　age of 19, 119, 120–1
　by children to sexual behaviour 126
　sexual offences against females and 120–1
　sexual offences against males and 119–20
coping strategies
　extreme trauma 76–7
　male co-survivors of rape 92
　sexually abused boys 43–4
co-survivors
　male, female rape victims 87–102
　psychological reactions 84
counselling
　male victims 11
　rape victims and their male co-survivors 92, 95–9

Index

stress inoculation training vs. 140
counsellors
  male, female rape victims 99–102
  training and supervision 135
  as victims 84
Criminal Law Revision Committee (1984) 23–4, 123
crisis intervention 137–8
  male co-survivors of rape 96
  rape survivors 99–100, 133, 134–5
cultural aspects, male sexual assault 104–14

depression 83
Deschauffours, Benjamin 111–12
desensitization, systematic 139
disclosure phase, sexual abuse 57
distraction strategies 76, 92, 93
dominance 8–9, 68, 80, 108, 113–14
drug abuse 7, 53, 70

education, male co-survivors of rape 96, 97–8
eighteenth century Europe 111–12
emotional processing
  completion 82–3
  everyday trauma 75–6
  extreme trauma and 76–7, 79
employment, homosexuals 17–18, 23
ethnology 112–14
Europe
  eighteenth century 111–12
  homosexual law reform 24–5
European Convention on Human Rights 24
exhibitionism 33
externalized responses, sexually abused boys 33, 43–4, 49–50, 52

families
  homosexuals 21–2
  sexually abused boys 56–8
father–son incest 38–9, 42

Gay Civil Liberties Trust 23
Gay London Policing Group (GALOP) 23
gender
  age of onset of sexual abuse 34
  assailants 1–2, 117, 120
  differential responses to trauma 132–3
  law on sexual offences and 122–3, 124–5
  responses to sexual abuse and 44–5, 52, 132

sexual abusers 29
sexual abuse victims 28–9
therapists 137
Gilles de Rais 109
Great Ormond Street child sexual abuse project 55–8
Greece 108
'Greek love' 17
gross indecency 19, 20, 117, 118
group treatment, victims of sexual abuse 55–8
guilt 81, 102

helplessness 8–9
hepatitis B 70
history, male sexual assault 104–14
HIV infection 2, 18–19, 70, 82
homicide, prisons 70
homophobia 10, 13–25, 34
  concept 13–14
  origins 14–15
homosexuality
  coercive, prisons and other institutions 68–9, 71–2, 73
  fear of, sexual abuse victims 33–4
  historical aspects 106–12
  sexually abused boys 42–3
  sexually assaulted men 9–10
homosexual panic 13–14
homosexuals
  activity in institutions 67, 70, 71
  attitudes to 17–18, 20–3
  child sexual abuse by 17
  law on sexual offences 118–20, 125–6
  legal discrimination against 19–20, 116–17, 124
  myths and half-truths 15–19
  official attitudes 23–5
  sexual assault by 1–2, 4, 9, 128
  sexual assaults on 6, 7–8, 20–1, 128–9
hospitals
  rape crisis intervention 133
  sexually abused children in 30
Howard League 23, 116, 124, 129
human immune deficiency virus (HIV) infection 2, 18–19, 70, 82
humiliation 68, 80

identification with aggressor, sexually abused boys 45
imaginal exposure, prolonged 140
importuning, male 19
incest
  father–son 38–9, 42
  law on 117, 118, 119, 120
  mother–child 28, 34–6

Index

sibling 36–7
under-reporting 34
indecency, gross 19, 20, 117, 118
indecent assault 5, 126
  criminal statistics 122–4
  law on 117, 118–19, 120, 121
  legal penalties for men vs. women 122–3, 124–5
initiation rites 72, 105–6
institutions
  homosexual activity 67, 70, 71
  male rape 67–73, 126
  sexual abuse 31
Islam 110

law on sexual offences 116–29
  against females 120–1
  against men 117–20
  discrimination against homosexuals 19–20, 116–117, 124
  in operation 122–4
  prospects of reform 23–5
Local Government Act, Section 28 (1988) 16, 24
London Lesbian and Gay Switchboard 23
London Rape Crisis Centre 134

male co-survivors of rape victims 87–102
  cognitive responses 89–92
  counselling 92, 95–9
  emotional and behavioural responses 92–5
male sexual assault (rape)
  community 1–11, 127
  cultural and historical aspects 104–14
  definitions in historical accounts 104–6
  institutions 67–73, 126
  legal discrimination against victims 129
  post-traumatic stress disorder and 79–81
  prevalence 1–2
  prisons 10, 67–72, 73, 126, 128
  treatment 7, 131–40
  under-reporting 128–9
marriage, rape in 121
masculinity
  dynamics, prevalence of sexual abuse and 33
  inappropriate attempts to reassert 43–5
media
  attitudes to homosexuals 22
  reporting male sexual assault 3
merchant navy 16
military institutions, sexual assault in 72–3

mother–child incest 28, 34–6
myths
  homosexuals 15–19
  rape 89–92
  sexual abuse 27

National Association of Victims' Support Schemes (NAVSS) 134
National Council for Civil Liberties 23

over-protection, rape victims by male partners 92–5, 98

paedophilia 17, 38–9, 47
parents
  homosexuals 21–2
  normal interactions with children 28
  sexual abuse by 28, 34–6, 38–9, 42
  sexual abuse victims 57–8
partners
  male, rape victims 87–102
  sexual assault/torture victims 84
passivity, *see* submission
perpetrators, *see* assailants
personality disorder, antisocial 52
personal relationships, rape victims and partners 88–9
personal vulnerability, post-traumatic stress disorder 78–9
physical abuse of children, sexual abuse and 38, 50
police 6, 20, 21, 23
post-traumatic stress disorder (PTSD) 77–81, 131–2
  depression and 83
  diagnostic criteria 77
  male co-survivors of rape 88
  personal vulnerability and 78–9
  sexual dysfunction 83
  sexually abused children 48
  sexual violence and 79–81
  treatment 82–3, 136
  see also rape trauma syndrome
prevalence
  male sexual assault 1–2
  prison rape 69–70
  sexual abuse 28–31
primates, sexual behaviour 113–14
prisons 10, 67–72, 73, 126, 128
  consequences of sexual assault 70–2
  nature and circumstances of sexual assault 68–9
  prevalence of sexual assault 69–70
promiscuity, sexual 8, 18, 48–9
prostitutes

child 30–1
male 16, 22
psychiatric disorders
  male victims 7, 10
  mothers abusing their sons 35–6
  sexually abused children 30, 51–2
psychodynamic psychotherapy 138
psychological reactions 75–85
  co-survivors of sexual trauma and torture 84
  female rape victims 10, 98, 127, 131
  gender differences 132–3
  male co-survivors of rape 87–102
  male rape victims 5–10, 13, 127–8
    immediate 5–6
    longer-term 6–7
  sexual abuse 33, 39–54, 131–2
  sexual assault in prisons and institutions 70–2
  see also post-traumatic stress disorder

'queer bashing' 9, 20

racial factors, prison rape 69
rape (female)
  historical aspects 106, 110–11
  law on 1, 120, 121
  male co-survivors 87–102
  male counsellors of survivors 99–102
  in marriage 121
  perceptions 126–7
  reactions 10, 98, 127, 131
  as shared life-crisis 87–9
  treatment interventions 135–40
  treatment providers 133–5
rape (male), *see* male sexual assault
rape crisis centres 11, 82, 131, 134–5
rape trauma syndrome 10, 88, 98, 135–6
recapitulation, sexually abused boys 45–7
rehabilitation phase, sexual abuse victims 58
religious attitudes, homosexuality 14–15, 22
residential schools 31
Romans, ancient 104–5, 106–8
runaways, sexual abuse 30

Sade, Marquis de 112
schools, residential 31
secondary victimization 82
self-esteem, victims of sexual abuse 53–4
separation, sexual abuse victims 57–8
sex rings, child 30–1
sexual abuse 27–58
  by children and adolescents 36–8
  definition 27–8
  effects 33, 39–54, 131–2
    general responses 40–1
    long-term 50–4
    specific responses 41–7
  extrafamilial 32–3
  father–son 38–9, 42
  Great Ormond Street treatment project 55–8
  by homosexuals 17
  identification of perpetrator risk in childhood 47–50
  indicators in boys 33, 34, 47–8
  intrafamilial 34
  management and treatment 54–8
  prevalence 28–31
  types 39
  under-reporting 31–9, 125
  victim–abuser cycle 45–7
  by women 34–6
sexual assault, male, *see* male sexual assault
sexual behaviour, age-inappropriate 33, 47–50
sexual dysfunction
  rape victims and their partners 89, 94–5
  sexual abuse victims 52–3
  sexual assault and torture victims 83
  treatment 140
sexual identity, sexually abused boys 42–3
sexual intercourse, unlawful 118, 120–1
sexualized attention 27–8
sexually transmitted diseases 1–2, 18–19, 70, 82
sexual offences, law on, *see* law on sexual offences
Sexual Offences Act (1967) 15, 20
Sexual Offences Act (1985) 123
Sexual Offences (Amendment) Act (1976) 1
sexual orientation
  assailants 1–2, 4, 9, 17
  long-term effects of sexual assault 7–8, 9–10
  male victims 4
  prison rape and 71–2
  sexually abused boys 42–3
sexual relationships
  male victims 7–8, 9
  rape victims and their partners 94–5
  sexual abuse victims 54
shell shock 78
sibling incest 36–7
sleep disturbances 78
sodomy 105, 109
stepfathers, sexual abuse by 38

stigma, male sexual assault 5–6, 10, 128, 129
Stonewall Housing Association 23
stress inoculation training 139–40
stuprum 106
submission (passivity) 9, 68, 104–5, 106–7, 113–14
substance abuse 52, 53
suicide/attempted suicide
  homosexuals 21–2
  male victims 7
  prisons 71
  sexual abuse victims 51
survivors 136
systematic desensitization 139

torture 75–85, 105
  definition 80
  depression after 83
  existential aspects 84
  post-traumatic stress disorder and 79–81
  sexual dysfunction after 83
trauma
  everyday, reactions to 75–6
  extreme, emotional processing and 76–7
  see also post-traumatic stress disorder
Trauma Symptom Checklist (TSC–33) 51–2
traumatic sexualization 41, 43

treatment
  male rape victims 7, 131–40
  post-traumatic stress disorder 82–3, 136
  providers 133–5
  reluctance to continue 137–8
  sexual abuse victims 54–8
  sexual dysfunction 140
  types 135–40
  volunteer organizations and 134–5

under-reporting 125
  male sexual assault 128–9
  prison rape 69–70
  sexual abuse 31–9, 125
United States of America, law on sexual offences 1, 24

victimization, secondary 82
victims 136
  male rape in community 4
  as perpetrators of sexual abuse 37, 45–50, 57
  prison rape 68–9, 71–2
  reactions, see psychological reactions
Victims Charter (Home Office 1990) 133–4
victim support schemes 11, 134–5
volunteer organizations 134–5